T0414210

Improving Soil Fertility Recommendations in Africa using the Decision Support System for Agrotechnology Transfer (DSSAT)

Job Kihara • Dougbedji Fatondji
James W. Jones • Gerrit Hoogenboom
Ramadjita Tabo • Andre Bationo
Editors

Improving Soil Fertility Recommendations in Africa using the Decision Support System for Agrotechnology Transfer (DSSAT)

Springer

Editors
Job Kihara
CIAT – Malawi Office
Tropical Soil Biology and Fertility (TSBF)
Institute of CIAT, c/o ICRAF, Nairobi
Kenya

James W. Jones
Agricultural and Biological Engineering
Department, University of Florida
Gainesville 32611, FL
USA

Ramadjita Tabo
Forum for Agricultural Research
in Africa (FARA), Accra
Ghana

Dougbedji Fatondji
Soil & water management
International Crop Research
Institute for the Semi-Arid Tropic
P.O. Box 12404
Niamey
Niger

Gerrit Hoogenboom
AgWeatherNet, Washington
State University, Prosser, WA 99350
USA

Andre Bationo
Soil Health Program
Alliance for a Green Revolution
in Africa (AGRA)
Agostino Neto Road, Airport Res. Area 6
Accra, PMB KIA 114, Airport-Accra
Ghana

ISBN 978-94-007-2959-9   ISBN 978-94-007-2960-5 (eBook)
DOI 10.1007/978-94-007-2960-5
Springer Dordrecht Heidelberg New York London

Library of Congress Control Number: 2012934565

Printed on acid-free paper

Springer is part of Springer Science+Business Media (www.springer.com)

Preface

In sub-Saharan Africa (SSA), increasing agricultural productivity is critical to meeting the food security and economic development objectives in the face of rapid population growth. Presently, the agricultural sector supports over 80% of the people in SSA, which is also the major contributor of GDP. A key challenge for scientists, governments and other stakeholders in the region is that food production should increase by 70% by the year 2050 to meet the caloric nutritional requirements of the growing population. Agricultural intensification is expected to be the main avenue for achieving these food increases. Crop models offer the benefit of increasing our understanding of crop responses to management in different soil and climatic conditions. Such responses are often of a complex and non-linear nature given the innumerable interactions among weather, soil, crop, and management factors throughout the growing season. Crop models can also provide insights in what might happen to productivity under various climate change scenarios, a domain beyond the reach of field experimentation. The outputs can inform key decision-makers at local, national, and regional levels in order to put the appropriate measures in place. Although major advances in modelling have been made in the USA, Europe and Asia, sub-Sahara Africa (SSA) lags behind due to the limited number of soil scientists and agronomists with the skills to set-up and run crop model simulations. Having a well-trained cadre of African modellers would greatly facilitate the design of best crop management and adaptation measures in the varied environments and to boost agricultural productivity in the region.

Over the past 20 years, efforts have been put in place to train scientists in the use of crop models, but the human resource base remains meagre. Most of the training was in the form of workshops and due to post-workshop financial constraints, limited or no follow-up efforts were made. Moreover, the disciplinary nature of university training in the region is not conducive to integrated, interdisciplinary, systems approaches. It is against this backdrop that the African Network for Soil Biology and Fertility (AfNet) and their collaborators, realizing that sustained follow-up was the key roadblock, organized a training programme which culminated in this publication. Many more such programmes are needed in order to strengthen the African modelling community in communicating effectively with decision makers as well as global community of modellers.

The chapters in this book present the context, key experiences and the results on the use of DSSAT in crop simulation. Chapter 1 presents the key steps and provides insights into building capacity for modeling in SSA. The experiences should inform capacity building efforts in order to choose carefully the training pathway. Chapter 2 summarizes the minimum data set required to set up and run crop models for (a) model applications, (b) general model evaluation and (c) detailed model calibration and evaluation. The chapter shows that little additional data could be all that one needs to have experimental data useful for modeling purposes. Chapter 3 discusses African soils and the key limitations to productivity. Chapter 4 focuses on sensitivities of DSSAT to uncertainties in input parameters while Chaps. 5–10 present key results of modelling from specific programs conducted in Ghana, Niger, Senegal and Kenya. The chapters present the key steps followed in the model calibrations and simulations for different themes including responses to fertilizer, organic resources and water management. Although the use of crop models is important in understanding African agriculture, there are key market and policy issues that must be addressed if agriculture is to be really improved. Thus Chap. 11 focuses on these issues and presents an integrated soil fertility management-innovative financing concept.

It is my hope that the approach to training, the model calibration and assessment procedures, the knowledge and wealth of experiences presented in this book will enhance the understanding and catalyse the use of crop growth models among the scientific community in Africa.

Prof. Dr. Paul L.G. Vlek
Executive director, WASCAL

Contents

Chapter 1
Building Capacity for Modeling in Africa

Andre Bationo, Ramadjita Tabo, Job Kihara, Gerrit Hoogenboom, Pierre C.S. Traore, Kenneth J. Boote, and James W. Jones

Abstract The use of models in decision support is important as field experiments provide empirical data on responses to only a small number of possible combinations of climate, soil, and management situations. Yet, crop modeling by African scientists so far has been limited. Therefore, to build the capacity of African scientists in the use of decision support systems, a provision was made for training within two main projects: Water Challenge Project (WCP) and Desert Margins Programme (DMP), jointly led by TSBF-CIAT (Tropical Soil Biology and Fertility Institute of the International Centre for Tropical Agriculture) and International Centre for Research in the Semiarid Tropics (ICRISAT). A unique approach to training on modeling was developed and was based on four main pillars: (a) learning by doing, (b) integrated follow-up, (c) continuous backstopping support and (d) multi-level training embedded in a series of three training workshops. Although crop models are useful they have limitations. For instance, they do not account for all of the factors in the field that may influence crop

A. Bationo (✉)
Soil Health Program, Alliance for a Green Revolution in Africa (AGRA) Agostino Neto Road, Airport Res. Area 6 Accra, PMB KIA 114, Airport-Accra, Ghana
e-mail: abationo@agra-alliance.org

R. Tabo
Forum for Agricultural Research in Africa (FARA), Accra, Ghana

J. Kihara
CIAT – Malawi Office, Tropical Soil Biology and Fertility (TSBF) Institute
of CIAT, c/o ICRAF, Nairobi, Kenya

G. Hoogenboom
AgWeatherNet, Washington State University, Prosser, WA 99350, USA

P.C.S. Traore
International Crops Research Institute for the Semi-Arid Tropics (ICRISAT), Bamako, Mali

K.J. Boote • J.W. Jones
Agricultural and Biological Engineering Department, University of Florida, Gainesville 32611, FL, USA

J. Kihara et al. (eds.), *Improving Soil Fertility Recommendations in Africa using the Decision Support System for Agrotechnology Transfer (DSSAT)*,
DOI 10.1007/978-94-007-2960-5_1, © Springer Science+Business Media Dordrecht 2012

yield and inputs must be accurate for simulated outputs to match observations from the field. Thus it is imperative that these issues are carefully considered and weighted before attempting to evaluate the predictability of a crop model. However, the use of crop models and decision support systems in concert with experiments can provide very useful alternative management options for resource-poor farmers in Africa and other regions across the globe.

Keywords Crop models • Decision Support Systems • Africa • Farmers • African scientists

Introduction

Farmers adapt their management systems to prevailing climate, soils, pests, and socio-economic conditions by selecting suitable crops, varieties, and management practices. Seasonal climate variability often results in highly variable yields that may cause economic losses, food shortages, inefficient resource use, and environmental degradation. Market and policy changes occur at the same time, thereby creating highly complex combinations of factors that farmers must consider when making decisions related to agricultural production. Information is needed to help farmers and policy makers to evaluate all these factors in order to anticipate changes and make decisions and policies that promote long-term sustainable management practices.

A major role of agricultural science is to develop methods for analyzing and selecting production options that are well adapted to the range of weather and climate conditions that may occur, taking into account the needs and capabilities of farmers in a given region. Crop responses to weather are highly complex and nonlinear; they are determined by many interactions among weather, soil, crop, and management factors throughout the growing season. Field experiments provide empirical data on responses to only a small number of possible combinations of climate, soil, and management situations. Also, existing management systems from other regions, new crops and varieties and other technologies being developed by scientists may provide useful adaptation options. However, it is impossible to conduct experiments that cover the full range of possible management options and climate conditions to determine production systems that are more resilient to climate variability, potential changes in climate, and farmers' goals (Nix 1984; Uehara and Tsuji 1991; Jones 1993). Instead of prescriptions, farmers need information on options that can increase their resilience and capacity to adapt to current climate risk and likely future climate conditions (Tsuji et al. 1998).

Nix (1984) criticized the predominance of a "trial and error" approach in agricultural research for evaluating management practices. He emphasized the need for a systems approach in which: (1) experiments are conducted over a range of environments; (2) a minimum set of data is collected in each experiment; (3) cropping system models are developed and evaluated; and (4) models are used to simulate production technologies under different weather and soil conditions so as to provide a broad range of potential solutions for farmers. Nix (1984) referred to the high cost of field experiments in addition to their limited extrapolation domain because results are site-specific. These concepts led to the development of the DSSAT (Decision Support System for Agrotechnology Transfer) under the auspices of the International Benchmark Sites Network for Agrotechnology Transfer (IBSNAT) Project suite of crop models that was designed to help researchers use this systems approach (e.g., IBSNAT 1989; Uehara and Tsuji 1991; Jones 1993; Jones 2003 Hoogenboom et al. 1994, 2004). Some crop simulation models and soil water models were already available (e.g., Ritchie 1972; de Wit and Goudriaan 1978; de Wit and Penning de Vries 1985; Jones et al. 1974; Williams et al. 1983; Arkin et al. 1976; Wilkerson et al. 1983), but prior to the IBSNAT initiative, there had not been a broad international effort focusing on the application of crop models to practical production situations. Although crop models were not originally developed for use in climate change research, they have been widely used for this purpose (e.g., Rosenzweig et al. 1995). They are well suited for these studies because they incorporate the effects of daily weather conditions on crop growth processes, predicting daily growth and development and ultimately crop yield. By simulating a crop grown in a particular soil, under specified management practices, and using a number of years of daily historical weather data at a site, one obtains an estimate of how a particular management system would perform under current and changed climate conditions.

The basic concept of crop modeling is that simulating crop growth and yield using dynamic crop models will produce results that represent how a real crop growing under specific environment and management conditions would perform. However, there are practical limitations that must be considered before making use of this approach in any study. One main limitation is that crop models do not account for all of the factors in the field that may influence crop yield. For example, crop diseases, weeds, and spatial variability of soils and management implementation can cause large differences in yield, and these factors are seldom included in crop simulation analyses. Another limitation is that inputs must be accurate or else simulated outputs are unlikely to match observations from the field. Attempts to evaluate the predictability of a crop model thus require that weather, management and soil inputs are measured in the field where the evaluation experiments are conducted. Furthermore, model evaluation experiments would ideally be designed to eliminate yield-reducing factors that are not included in the model. And finally, parameters that are used to model the dynamics of soil and crop processes need to be accurate for comparison with observed field data. For example, if one uses a crop model to simulate crop yield responses to water or N management using incorrect soil water parameters, results will show that the model fails to mimic results from field experiments or, more problematically, provide results that may mislead researchers or other model users.

Capacity Building

The use of models in decision support by African scientists is limited. Although most research on land productivity has traditionally focused on plot level approach, there has been low extrapolation of the findings to wider scales. The main problem is the limited availability of agricultural scientists (both soil scientists and agronomists) due to low resource allocation to training and capacity building in

African countries (Bationo et al. 2004). Secondly, the training approach employed in most training institutions especially those of higher learning in Africa is disciplinary. Modeling for extrapolation requires integration of various disciplines in what is now called systems approach and is based on the practical impossibility to do research everywhere.

In order to build capacity of African scientists in use of decision support systems, a provision was made for training within two main projects, Water Challenge Project (WCP) and Desert Margins Programme (DMP), undertaken jointly by TSBF-CIAT and ICRISAT among other partners. WCP aimed to enhance water productivity through the integration of water efficient and high yielding germplasm, water and soil conservation options, and nutrient management technologies coupled with strategies for empowering farmers to identify market opportunities, and scaling up appropriate technologies, methodologies and approaches. The project was implemented in Burkina Faso, Niger and Ghana. The specific objectives were to:

1. Develop, evaluate and adapt, in partnership with farmers, integrated technology options that improve water and nutrient use efficiency and increase crop yields in the Volta Basin.
2. Develop and evaluate methodologies, approaches and modern tools (GIS, models, farmer participatory approaches) for evaluating and promoting promising water, nutrient and crop management technology options.
3. Improve market opportunities for small holder farmers and pastoralists, identify and assess market institutional innovations that provide incentives for the adoption of improved water, nutrient and crop management technologies that benefit different categories of farmers, especially women and other marginalized groups of farmers.
4. Build the capacities of farmers and rural communities to make effective demands to research and development organizations, and influence policies that promote the adoption of sustainable water and nutrient use technologies.
5. Promote and scale up and out 'best bet' crop, water, and nutrient management strategies in the Volta Basin through more efficient information and methodology dissemination mechanisms.

Desert Margins Program (DMP) initiated in 2003 under the funding of UNEP-GEF operated in nine African countries namely: Burkina Faso, Botswana, Mali, Namibia, Niger, Senegal, Kenya, South Africa, and Zimbabwe. The overall objective of the DMP was to arrest land degradation in Africa's desert margins through demonstration and capacity building activities developed through unravelling the complex causative factors of desertification, both climatic (internal) and human-induced (external), and the formulation and piloting of appropriate holistic solutions. The project addressed issues of global environmental importance, in addition to the issues of national economic and environmental importance, and in particular the loss of biological diversity, reduced sequestration of carbon, and increased soil erosion and sedimentation. Key sites harbouring globally significant ecosystems and threatened biodiversity serve as field laboratories for demonstration activities related to monitoring and evaluation of biodiversity status, testing of most promising natural resources options, developing sustainable alternative livelihoods and policy guidelines and replicating successful models. In this project, models serve as decision guides for extrapolation of field results to wider recommendation domains. The broader objectives of the overall DMP were to:

1. Develop a better understanding of the causes, extent, severity and physical processes of land degradation in traditional crop, tree, and livestock production systems in the desert margins, and the impact, relative importance, and relationship between natural and human factors.
2. Document and evaluate, with the participation of farmers, NGO's, and NARS, current indigenous soil, water, nutrient, vegetation, and livestock management practices for arresting land degradation and to identify socio-economic constraints to the adoption of improved management practices.
3. Develop and foster improved and integrated soil, water, nutrient, vegetation, and livestock management technologies and policies to achieve greater productivity of crops, trees, and animals to enhance food security, income generation, and ecosystem resilience in the desert margins.
4. Evaluate the impact and assist in designing policies, programs, and institutional options that influence the incentives for farmers and communities to adopt improved resource management practices.
5. Promote more efficient drought-management policies and strategies.
6. Enhance the institutional capacity of countries participating in the DMP to undertake land degradation research and the extension of improved technologies, with particular regard to multidisciplinary and participative socio-economic research.
7. Facilitate the exchange of technologies and information among farmers, communities, scientists, development practitioners, and policymakers.
8. Use climate change scenarios to predict shifts in resource base and incorporate these into land use planning strategies.

Within the framework of these two main projects, we identified the need for new scientific and technical training on the use of DSSAT models in order to hasten implementation and fulfillment of all the proposed outputs.

A New Approach

We developed a unique approach to modeling training based on four main pillars: (1) learning by doing, (2) integrated follow-up, (3) continuous backstopping support and (4) multi-level training. Our learning by doing strategy required that scientists being trained not only work on individual computers for hands-on-experience but also collect their own data that was used to run the models. Data collection by the scientists was done within the framework of the two main projects (WCP and DMP) as well as in the African Network for soil biology and fertility (AfNet of TSBF-CIAT) supported sites. The arrangement attracted self-sponsored scientists working in Africa in addition to those financed through the two projects. Follow-up was achieved through continuous communication of the organizers who were also the lead investigators within WCP and DMP and the scientists using data from these projects. A minimum dataset for DSSAT was developed for use by scientists as a checklist during field data collection. A concise summary of data requirements for modelling is presented in Hoogenboom et al. (2012, this volume). Professional and technical backstopping support was given by scientists associated with the International Consortium for Agricultural Systems Applications (ICASA) and progressive DSSAT modelers working in Africa mainly ICRISAT and IFDC. Scientists and organizers were continuously in contact with the trainers during and after a training workshop. Modeling is quite complex and one training session often does not lead to sufficient understanding and know-how for use of models. TSBF-CIAT and ICRISAT-Niamey in conjunction with ICASA therefore organized a series of three workshops. The training workshops focused on both biophysical and socio-economic issues to allow the screening and identification of scenarios that will lead to best bet management practices and policies for rebuilding biodiversity and restoring degraded and collapsed ecosystems.

The first workshop, held in Arusha Tanzania in 2004, was to expose people to the theory and familiarize with DSSAT software and its operations as well as on general modeling concepts. The second workshop, held in Accra Ghana in 2005, aimed at enabling trainees to input and use their own datasets in DSSAT as well as familiarize them with the minimum dataset concept for modeling. The scientists then used the period 2005–2007 to collect the required minimum dataset and or fill in gaps in the data they already held. Thus, the third training and last in the series was held in Mombasa Kenya in 2007 to have the trainees model different scenarios using their own datasets and write a scientific manuscript for publication. The training workshops provided participants, mainly young scientists with an opportunity to learn from model developers, to peer review and positive criticism and information sharing between sub-regions and countries.

The themes addressed by scientists include: tillage and nitrogen applications, soil and water conservation practices including effects of zai technology, phosphorus and maize productivity, generation of genetic coefficients, long-term soil fertility management technologies in the drylands, microdosing, manure and nitrogen interactions in drylands, optimization of nitrogen x germplasms x water, spatial analysis of water and nutrient use efficiencies, and tradeoff analysis.

Conclusions

Crop models are useful for simulating crop and soil processes in response to variations in climate and management. Building a critical mass of African modelers requires an integrated approach to learning at the start of a scientific career. Training of scientists in crop modeling should be step-wise and systematic to ensure the scientists gain the minimum ability to start using models. A minimum dataset of good quality is required to ensure accurate comparison with observed field data. Attempts to evaluate the predictability of a crop model require that whenever possible, weather, management and soil inputs are measured in the field where the evaluation experiments are conducted. Crop models should be evaluated with caution as they seldomly contain all of the factors in the field that may influence crop yield, e.g., crop diseases, weeds, and spatial variability of soils and management implementation that can cause large differences in yield.

References

Arkin GF, Vanderlip RL, Ritchie JT (1976) A dynamic grain sorghum growth model. Trans ASAE 19:622–630

Bationo A, Kimetu J, Ikerra S, Kimani S, Mugendi D, Odendo M, Silver M, Swift MJ, Sanginga N (2004) The African network for soil biology and fertility: new challenges and opportunities. In: Bationo A (ed) Managing nutrient cycles to sustain soil fertility in Sub-Saharan Africa. Academy Science, Nairobi

de Wit CT, Goudriaan J (1978) Simulation of ecological processes. A Halsted Press Book, Wiley, New York, 175 pp

de Wit CT, Penning de Vries FWT (1985) Predictive models in agricultural production. Philos Trans R Soc Lond B 310:309–315

Hoogenboom G, White JW, Jones JW, Boote KJ (1994) BEANGRO: a process-oriented drybean model with a versatile user-interface. Agron J 86:182–190

Hoogenboom G, Jones JW, Wilkens PW, Porter CH, Batchelor WD, Hunt LA, Boote KJ, Singh U, Uryasev O, Bowen WT, Gijsman AJ, du Toit A, White JW, Tsuji GY (2004) Decision support system for agrotechnology transfer version 4.0 [CD-ROM]. University of Hawaii, Honolulu

IBSNAT (1989) Decision support system for agrotechnology transfer v 2.1 (DSSAT v2.1). Department of Agronomy and Soil Science, University of Hawaii, Honolulu

Jones JW (1993) Decision support systems for agricultural development. In: Penning de Vries F, Teng P, Metselaar K (eds.) Systems approaches for agricultural development. Kluwer Academic, Boston, pp 459–471

Jones JW, Hesketh JD, Kamprath EJ, Bowen HD (1974) Development of a nitrogen balance for cotton growth models – a first approximation. Crop Sci 14:541–546

Jones JW, Hoogenboom G, Porter CH, Boote KJ, Batchelor WD, Hunt LA, Wilkens PW, Singh U, Gijsman AJ, Ritchie JT (2003) The DSSAT cropping system model. Eur J Agron 18:235–265

Nix HA (1984) Minimum data sets for agrotechnology transfer. In: Proceedings of the international symposium on minimum data sets for agrotechnology transfer, ICRISAT, Patancheru, Andhra Pradesh, India, 1984, pp 181–188

Ritchie JR (1972) Model for predicting evaporation from a row crop with incomplete cover. Water Resour Res 8:1204–1213

Rosenzweig C, Allen LH Jr, JW Jones, Tsuji GY, Hildebrand P (eds) (1995). Climate change and agriculture: analysis of potential international impacts. ASA Special Publication No. 59, Amer Soc Agron, Madison, 382 pp

Tsuji GY, Hoogenboom G, Thornton PK (eds) (1998) Understanding options for agricultural production, Systems approaches for sustainable agricultural development. Kluwer, Dordrecht, 400 pp. ISBN 07923-4833-8

Uehara G, Tsuji GY (1991) Progress in crop modeling in the IBSNAT project. In: Muchow RC, Bellamy JA (eds) Climatic risk in crop production: models and management for the semiarid tropics and subtropics. CAB International, Wallingford, pp 143–156

Wilkerson GG, Jones JW, Boote KJ, Ingram KT, Mishoe JW (1983) Modeling soybean growth for crop management. Trans ASAE 26:63–73

Williams JR, Jones CA, Dyke PT (1983) A modeling approach to determining the relationship between erosion and productivity. Trans ASAE 27:129–144

Chapter 2
Experiments and Data for Model Evaluation and Application

Gerrit Hoogenboom, James W. Jones, Pierre C.S. Traore, and Kenneth J. Boote

Abstract Crop models and decision support systems can be very useful tools for scientists, extension educators, teachers, planners and policy makers to help with the evaluation of alternative management practices. Many of the current crop models respond to differences in local weather conditions, soil characteristics, crop management practices and genetics. However, computer-based tools require inputs in order to provide reliable results. Especially for those new to crop modeling, the data requirements are sometimes somewhat overwhelming. In this chapter we provide a clear and concise summary of the input data requirements for crop modeling. We differentiate between requirements for model evaluation, model application and model development and improvement. For model inputs we define daily weather data, soil surface and profile characteristics, and crop management. For model evaluation and improvement we define crop performance data as it relates to growth, development, yield and yield components, as well as additional observations. We expect that this chapter will make the use and application of crop models and decision support systems easier for beginning modelers as well as for the more advanced users.

Keywords Crop modeling • Simulation • Decision support systems • Minimum data set • DSSAT • Cropping System Model (DSSAT) • CROPGRO• CERES

G. Hoogenboom (✉)
AgWeatherNet, Washington State University, Prosser, Washington 99350, USA
e-mail: gerrit.hoogenboom@wsu.edu

J.W. Jones
Department of Agricultural and Biological Engineering, University of Florida, Gainesville, Florida 32611, USA

P.C.S. Traore
International Crops Research Institute for the Semi-Arid Tropics (ICRISAT), Bamako, Mali

K.J. Boote
Department of Agronomy, University of Florida, Gainesville, Florida 32611, USA

Introduction

With the increasing interest in the applications of crop modeling and decision support systems, there is a need to clearly define the type of experiments that are required for both crop model evaluation and application. Especially for those new to crop modeling it is unclear what types of experiments should be conducted and what information should be collected in these experiments. Over the years several publications have been written to document these requirements (IBSNAT 1988; Hunt and Boote 1998; Hunt et al. 2001). The most extensive ones can be found in the documentation that was developed for the Decision Support System for Agrotechnology Transfer (DSSAT) Version 3.5, especially Volume 4 (Hoogenboom et al. 1999). This information is still relevant and has been included as electronic documents in the documentation section of DSSAT Version 4.0 (Hoogenboom et al. 2004) and DSSAT v4.5 (Hoogenboom et al. 2010).

Volume 4.8 entitled "Field and Laboratory Methods for the Collection of the Minimum Data Set" by Ogoshi et al. (1999) is based on Technical Report 1 that was published by the International Benchmark Sites Network for Agrotechnology Transfer (IBSNAT) Project (IBSNAT 1988). It includes extensive documentation on data collection procedures for modeling. In volume 4.7 entitled "Data Requirements for Model Evaluation and Techniques for Sampling Crop Growth and Development" Boote (1999) provides detailed procedures on the actual sampling techniques for growth analysis and crop development. However, an easy to use summary is currently not available. The goal of this chapter is, therefore to provide a clear and concise summary for experimental data collection for model evaluation and application.

Overview

In order to run a crop model and to conduct a simulation, a set of data are required. Sometimes this is referred to as a "Minimum Data Set." The terminology Minimum Data Set was first introduced by the IBSNAT Project. Although the type and details required for model inputs might vary somewhat depending on the crop or agricultural model, in general we can differentiate between three broad levels or groups. Level 1 defines the data required for model applications, Level 2 defines the data required for general model evaluation, and Level 3 defines the data required for detailed model calibration and evaluation. Potentially this type of data can also be used for the development of a model for a crop for which currently no dynamic crop simulation model exists.

Level 1 includes daily weather data, soil surface characteristics and soil profile information, and crop management. Level 2 includes the environmental and management data from Level 1 and some type of observational data that are collected during the course of an experiment. At a minimum the two key phenological phases, i.e., flowering or anthesis and physiological or harvest maturity, and yield and yield components are needed for observational data. Level 3 would include the environmental, management and observational data described under Level 2 and additional observations related to growth and development, such as growth analysis, soil moisture content, and soil and plant nitrogen, phosphorus, potassium, and others, depending on the overall intended model application or evaluation.

Experiments and Modeling

It is important to understand that one rarely develops an experiment for modeling only, but that experiments should be conducted in such a manner that they also have a modeling component that can be used for either model evaluation or application or both. It is also important to keep in mind that some of the basic data that are required for any model application, especially those described under Level 1, should be a basic set of data that are collected for documentation of any experiment. For instance, for many experiments local weather and soil conditions have a major impact on the outcomes of an experiment and should be included as part of the overall analysis.

Location of Experiments

Normally data for model evaluation are obtained from experiments, although in some cases one might only have access to statistical yield and production data. Although this information can be used, one should understand the level of detail and the quality of this type of data and expected outcomes with respect to the accuracy of the evaluation of a model. In general experiments can be conducted under controlled management conditions, referred to as "on-station" and in farmers' fields, referred to as "on-farm." For Level 3 one normally would not use data from on-farm experiments, but the data can be useful for Level 2 model evaluation if one understands the limitations of the data, such as the lack of replications in most cases, variability of environmental conditions and uncertainty of the inputs. In some cases experiments can be conducted in growth chambers or in Soil-Plant-Atmosphere Research (SPAR) chambers where most environmental conditions can be controlled. However, for accurate model evaluation, on-station experiments with at least three or four replications are preferred.

General Purposes of Experiments

It is always important to keep the overall goal of the research in mind and design appropriate experiments accordingly, rather than concentrating on the model only. There is a wide range of applications with some of the key ones listed below.

– Technology evaluation, such as evaluation of new cultivars, inputs, including irrigation and fertilizers, and soil preparation, such as tillage and conservation agriculture.

- Characterization of yield limiting factors in order to focus on new technology development and evaluation.
- Understanding the interactions among management factors, such as water, nutrients, etc., and aiming at refining agricultural management technologies.
- Understanding the interactions of the environment, such as increases in temperature and CO_2.
- Understanding the interactions between genotype and environment (G x E).
- Long-term soil sustainability and soil health, including improvement of soil organic matter.
- Understanding environmental impact, such as nitrogen pollution due to different management practices.
- Potential application of agricultural crops for food, feed, fiber and fuel production.

General Purposes of Model Use

It is important to determine the overall purpose of the use of modeling and how it contributes to the overall research goal. In many cases adding a systems analysis and modeling component can strengthen the overall research approach. A partial list of model applications is listed below.

- Understand and interpret experimental results.
- Enhance quality of field research and the results that are derived from it.
- Diagnose yield gaps by looking at the differences between potential, attainable and actual yield from on-station and on-farm research, and to help develop technologies to test these under field conditions.
- Help publish results of field trials via systems and modeling analysis.
- Estimate impacts on production, water use, nitrogen use, and other inputs and determine various resource use efficiencies at scales from field to farm to watershed to region and higher.
- Estimate economic implications of different technologies.
- Estimate the impact of climate change and climate variability on crop production and develop adaptation scenarios.
- Plant breeding, Genotype * Environment interaction and the development of ideotypes.
- Enhance interdisciplinary research through interaction of soil scientists, agronomists, economists, engineers, GIS/remote sensing scientists, and others.

Level 1 Data

Level 1 data for model applications include daily weather data, soil characteristics and crop management. These data are an absolute requirement for any successful model evaluation and application. Well-documented experiments normally already include this information or researchers can provide you with the source of the data.

Weather Data Required (Daily)

1. <u>Minimum and maximum temperature</u>
2. <u>Precipitation or rainfall</u>
3. <u>Total solar radiation or sunshine hours</u>
4. <u>Dewpoint temperature or relative humidity</u>. If an automated weather station is used for weather data collection, this information is normally readily available
5. <u>Average daily wind speed or daily wind run</u>. Again, if an automated weather station is used for weather data collection, this information is normally readily available

Measured or recorded minimum and maximum temperature and rainfall have to be measured with a gauge, thermometer or environmental sensor, while in some cases other methods can be used to help estimate daily solar radiation based on satellites (White et al. 2011) or simple equations or solar radiation generators (Garcia y Garcia and Hoogenboom 2005).

It would be important to the overall value of an experimental data set to have an automatic weather station collect daily weather data in areas where experiments are to be conducted. In addition, rainfall is needed for all sites where experiments are performed, including on-farm as well as on-station due to its significant spatial and temporal variability. Otherwise, it will not be possible to interpret crop yield response or to understand the interactions of water, nutrients and management. Temperature and solar radiation could be measured in the areas where several experiments are conducted. Generally, temperatures and solar radiation may change little over distances of about 50 km, but this is not true near the coast or when elevation changes occur in the area. One could situate a weather station in the center of an area where experiments will be conducted either on-farm or on-station.

Soil Data

Soil information includes general site and soil surface information and soil profile characteristics.

1. <u>General site information</u>

   - Latitude, longitude and elevation

2. <u>Soil surface information</u>

   - Soil taxonomy (if available)
   - Soil slope

- Soil color
- Stones (%)

3. Soil profile data, for each soil horizon in which roots are likely to grow

    - Soil texture, including % sand, silt, and clay and stones, especially for the surface layers
    - Soil organic carbon
    - Bulk density is desirable
    - Lower Limit of plant extractable soil water (LL) or permanent wilting point and Drained Upper Limit (DUL) or field capacity. Field measurements are desirable, but there are various methodologies to help estimate these parameters

Initial Conditions

When running the model with the soil water, nitrogen, phosphorus, organic carbon and other soil components turned on, it is important to define the initial conditions of the soil profile at the start of the simulation. This might not necessarily be the planting date. If initial conditions are not defined, the model normally defaults to a 0 value for all soil nutrients, while soil moisture is set to the drained upper limit. We realize that for many scientists it is rather difficult to measure these initial conditions, or they do not have the resources. DSSAT, therefore, has tools to help you estimate values. Note that the surface and soil residues are the residues of the previous crop only. Any supplemental residues or manure are considered part of the inputs.

1. Previous field history

    - For the simulation of the soil organic carbon balance a user has to know the previous crop history.

2. Initial soil profiles conditions

    - Initial soil moisture versus depth
    - Initial nutrients (NO_3, NH_4, P) versus depth
    - Other soil chemical properties as needed for the experimental objectives

3. Surface residues at the start of simulation or at planting

    - Crop type or manure type
    - Total amount as dry weight
    - %N and %C (and %P) contents
    - Incorporation depth and % incorporation
    - Note that if resides or manure are applied at planting they are considered to be part of crop management

Management Data

Crop management data seems like a relatively easy set of information to obtain. However, it is surprising to find that many experiments lack detailed information except if it is part of a treatment. For instance, if supplemental irrigation is applied but it is not part of the overall set of treatments, it is sometimes difficult to obtain the actual dates of the irrigation applications and the amount of water that was applied.

1. Planting

   • Date of planting
   • Plant spacing or density. This would be the official plant stand and not the seeding or sowing density.
   • Crop and cultivar name and its characteristics.
   • Planting material, e.g. seed, stick, etc.
   • Planting mode, e.g., row, hill, flat, ridge, etc.

2. Input information

   • Irrigation amount and the timing of the irrigation application
   • Fertilizer amount and type, timing of the fertilizer application, placement depth and application method
   • Amount of organic manure or residue, composition, time of the application, placement or incorporation depth and method of application
   • Amount and type of chemicals applied and for what purposes

Level 2 Data

Level 2 data includes the basic input data defined under level 2, including weather conditions, soil characteristics, crop management and initial conditions. In addition it includes some type of observational data in response to differences among treatments. This could range from different environments, including soil and weather conditions, different crops, different cultivars, varieties or hybrids, or different inputs.

Crop and Soil Response Measurements

1. Treatments. Although this seems obvious, it is important to understand the different treatment factors and their associated levels.

2. Yield and yield components

- Grain yield (kg/ha)
- Anthesis or flowering date and maturity date. Time of first seed or first grain would also be helpful
- Number of main stem nodes
- Above ground biomass, excluding grain
- Plant density at harvest
- Number of ears, pods, or other fruiting structures per unit area
- Average weight per unit grain, seed, fruit or other harvested material
- N and P concentrations of grain and other plant components

3. General observations

- Weeds and weed management. It is important to document if the weeds affected the actual outcome of the experiment, such as yield and biomass
- Pests and disease occurrence, including the date of the infection intensity, and actual damage
- Damage due to extreme weather events, such as hail, rainstorms, wind gusts, etc.
- General health of the crop

Level 3 Data

There are two types of data that provide crucial information related to growth response and water and nutrient use dynamics. These are detailed crop response measurements and detailed soil response measurements. However, they are expensive and time consuming, so it is impractical to collect these data in all experiments, especially if the experiments are conducted in farmers' fields. To obtain the data from these measurements in a few places would greatly increase the confidence in the use of models and the simulation of the responses in fields where they are not measured. In particular, water use by crops is highly critical to yield formation and is important in any technology aimed at increasing productivity and sustainability of soils in rainfed agriculture. The models predict root growth and water uptake, but credibility of those predictions will require good inputs as well as evaluation in soils and cropping systems in the region in which they are to be applied. Thus, the following measurements are suggested.

1. Growth analysis measurements. These measurements should be taken at least four to eight times during the growing season at regular spaced intervals, such as weekly or bi-weekly and one detailed sample at final harvest. Previous recommendations were to collect growth analysis samples at critical growth stages, but we have changed this approach and now prefer regularly spaced intervals for growth analysis sampling. Generally, these samples would be taken in on-station experiments only, and for some experiments only. This is mainly to make sure that you have good genetic coefficients and can simulate attainable yield or potential yield

- Above ground biomass
- Leaf, stem, seed, and ear (pod) mass
- Number of main stem nodes or leaves
- Leaf area index, if possible either destructively or in the field using a hand-held device
- N and P concentrations in plant parts
- Numbers of ears (pods) per square meter

2. Soil water content versus depth. There are various measurement techniques including gravimetric samples, time domain reflectometry (TDR), neutron probes and other electronic sensors. Samples or readings should be taken at least weekly or preferably more often. Moisture sampling after a prolonged wet period can be useful to determine the upper limits of water storage for the respective soil layers. Moisture sampling following a well-grown crop experiencing terminal drying is useful for determining the crop lower limits for water extraction.

3. Soil fertility versus depth

- Analysis for NO_3-N (ppm) for soil layers and sampling times are similar to what was discussed for soil water. However, these measurements can be taken less frequently due to the additional resources required and costs associated with sample collection and laboratory analysis.
- Soil Organic Carbon (%) for soil layers to rooting depth. At least one entire profile should be sampled for each soil type. This information is extremely critical, especially for eroded soils commonly found in the tropics. For other sites with the same soil type, a surface layer sample will suffice. In this case it is best to make the layer thickness applicable to the un-sampled profile. Total soil N could be useful to help determine the overall C:N ratio.
- Olsen-P, Total P and Org P for the top two soil layers should suffice.

Summary

This chapter provides a detailed review of the environmental and crop management data required for model simulation and the crop and soil response data for model evaluation. The DSSAT approach has been to emphasize model evaluation with locally available data to demonstrate that the model is able to simulate management and environmental responses and to determine local genetic coefficients. This is important, not only for a researcher, but also for those who will be using the outcomes of the simulation studies for policy and planning. Although the list of data might seem overwhelming for a novice modeler, most of the data presented here are in some form already available, especially for well-managed on-station experiments. Prior to designing and developing an experiment for model evaluation, make sure that this experiment has not already been conducted by a colleague at your university or a sister institution in the region. If you have to conduct an experiment, make sure to partner with others to compliment your skills and expertise. At the end the results will be much better and the outcomes of the modeling study much stronger. If questions still remain, make sure to use the resources that are available through international networks of model users, such as the DSSAT group (www.DSSAT.net).

References

Boote KJ (1999) Data required for model evaluation and techniques for sampling crop growth and development. DSSAT v3, vols 4–7. University of Hawaii, Honolulu, pp 201–216

Garcia y Garcia A, Hoogenboom G (2005) Evaluation of an improved daily solar radiation generator for the southeastern USA. Climate Res 29:91–102

Hoogenboom G, Wilkens PW, Tsuji GY (1999) DSSAT v3, vol 4. University of Hawaii, Honolulu, 286 pp. ISBN 1-886684-04-9

Hoogenboom G, Jones JW, Wilkens PW, Porter CH, Batchelor WD, Hunt LA, Boote KJ, Singh U, Uryasev O, Bowen WT, Gijsman AJ, du Toit A, White JW, Tsuji GY (2004) Decision support system for agrotechnology transfer version 4.0. [CD-ROM]. University of Hawaii, Honolulu

Hoogenboom G, Jones JW, Wilkens PW, Porter CH, Boote KJ, Hunt LA, Singh U, Lizaso JL, White JW, Uryasev O, Royce FS, Ogoshi R, Gijsman AJ, Tsuji GY (2010) Decision support system for agrotechnology transfer version 4.5. [CD-ROM]. University of Hawaii, Honolulu

Hunt LA, Boote KJ (1998) Data for model operation, calibration, and evaluation. In: Tsuji GY, Hoogenboom G, Thornton PK (eds) Understanding options for agricultural production. Systems approaches for sustainable agricultural development. Kluwer Academic, Dordrecht

Hunt LA, White JW, Hoogenboom G (2001) Agronomic data: advances in documentation and protocols for exchange and use. Agr Syst 70:477–492

IBSNAT (1988) Technical report 1, experimental design and data collection procedures for IBSNAT, 3rd edn revised. Department of Agronomy and Soil Science, College of Tropical Agriculture and Human Resources, University of Hawaii, Honolulu

Ogoshi RM, Cagauan BG Jr, Tsuji GY (1999) Field and laboratory methods for the collection of the minimum data set. DSSAT v3, vols 4–8, pp 217–286. University of Hawaii, Honolulu, Hawaii, 96822

White JW, Hoogenboom G, Wilkens PW, Stackhouse PW, Hoell JM (2011) Evaluation of satellite-based, modeled-derived daily solar radiation data for the continental United States. Agron J 103(4):1242–1251

Chapter 3
Knowing the African Soils to Improve Fertilizer Recommendations

Andre Bationo, Alfred Hartemink, Obed Lungu, Mustapha Naimi, Peter Okoth, Eric Smaling, Lamourdia Thiombiano, and Boaz Waswa

Abstract Africa is the continent with the lowest fertilizer use per hectare notwithstanding the fact it possesses geologically old, infertile and degraded soils. This chapter discusses the agro-ecological zones (AEZ) and main soil types in Africa followed by a section on the extent, effects and costs of land degradation including issues of soil productivity and profitability associated with fertilizer use in Africa. There are a variety of soil types in the five major agro-ecological zones of Africa. Ferralsols and the Acrisol are dominant in the humid zones. Ferralsols are dominant in the sub-humid zone and so are Lixisols while in the semi-arid zone Lixisols have

A. Bationo (✉)
Soil Health Program Alliance for a Green Revolution in Africa (AGRA)
Agostino Neto Road, Airport Res. Area 6 Accra, PMB KIA 114, Airport-Accra, Ghana
e-mail: abationo@agra-alliance.org

A. Hartemink
University of Wisconsin – Madison, Department of Soil Science,
1525 Observatory Drive, Madison, WI 53706-1299, USA
e-mail: hartemink@wisc.edu

O. Lungu
Department of Soil Science, University of Zambia, P.O. Box 32379, Lusaka, Zambia
e-mail: lunguoi@yahoo.co.uk

M. Naimi
Centre Régional de la Recherche Agronomique,
BP 589 Route Tertiaire Tnine Ouled Bouziri Aïn Nezagh, 26000 Settat, Morocco
e-mail: mnaimi@ifdc.org

P. Okoth
Tropical Soil Biology and Fertility institute of CIAT, P.O. Box 30677-00100, Nairobi, Kenya
e-mail: p.okoth@cgiar.org

E. Smaling
Group Soil Mapping and Land Evaluation, Environmental Sciences Group, Wageningen
University and Research Centre, Wageningen, the Netherlands
e-mail: esmaling@tiscali.nl

J. Kihara et al. (eds.), *Improving Soil Fertility Recommendations in Africa using the Decision Support System for Agrotechnology Transfer (DSSAT)*,
DOI 10.1007/978-94-007-2960-5_3, © Springer Science+Business Media Dordrecht 2012

the larger share. Sixty five percent of the agricultural land in Africa is degraded and soil fertility depletion, a manifestation of soil degradation, is currently a serious threat to food security among small-holder farmers. Because of this state of affairs there is a strong case for enhanced fertilizer use. Maize yield has reportedly increased over the control due to NPK fertilizer application from various AEZ and when soils are amended with lime and manure yield response has been even higher. Indeed, there is credible evidence of fertilizer response and profitability in Africa relative to other parts of the world, particularly, for maize and rice thus making fertilizer investment worthwhile. In conclusion, there is need for agricultural intensification through efficient use of soil nutrient and water resources. Technologies need to be adapted to the specific bio-physical and socio-economics circumstances of the small scale farmers in Africa. There is also need to focus more on increasing the fertilizers use efficiency and the development of the local fertilizer sector in order to make them more profitable.

Introduction

Africa covers an area of about 3.01×10^9 ha, out of which about 230×10^6 ha represents natural water resources (FAO 1978). Most soils of Africa are poor compared to most other parts of the world. Lack of volcanic rejuvenation has caused the continent to undergo various cycles of weathering, erosion and leaching, leaving soils poor in nutrients (Smaling 1995). Uncultivated soils have a natural fertility determined by soil-forming factors such as parent material, climate, and hydrology. For soils under natural vegetation, there is a virtual equilibrium but as soon as the land is altered through clearing of the natural forest or savanna, this equilibrium is broken and soil fertility declines at a rate depending on the intensity of cropping and replacement of nutrient loss in the systems (Smaling 1995). In addition to low inherent fertility, African soils nutrient balances are often negative indicating that farmers mine their soils. During the last 30 years, soil fertility depletion has been estimated at an average of 660 kg $N ha^{-1}$, 75 kg $P ha^{-1}$ and 450 kg $K ha^{-1}$ from about 200 million hectares of cultivated land in 37 African countries. Africa loses \$4 billion per year due to soil nutrient mining (Smaling 1993). The annual (1993–1995) use of nutrients in Africa averages about 10 kg of NPK per hectare. Fertilizer use ranges from nearly 234 kg per hectare in Egypt to 46 kg in Kenya to less than 10 kg in most countries in Sub-Saharan Africa. North Africa, with 20% of the continent's land area, accounts for 41% of the fertilizer consumption (Henao and Baanante 1999a). Whereas in the developed world, excess applications of fertilizer and manure has damaged the environment, the low use of inorganic fertilizer is one of the main causes for environmental degradation in Africa. Increased inorganic

L. Thiombiano
FAO Regional Office in Accra, Accra, Ghana
e-mail: Lamourdia.Thiombiano@fao.org

B. Waswa
Zentrum fur entwicklungsforsung (ZEF), Bonn, Germany
e-mail: bswaswa@yahoo.com

fertilizer use benefits the environment by reducing the pressure to convert forests and other fragile lands to agricultural uses and, by increasing biomass production, help increase soil organic matter content (Wallace and Knausenberger 1997).

As a result of the inherent low fertility of African soils and subsequent land degradation, only 16% of the land has soil of high quality and about 13% has soil of medium quality. About 9 million km^2 of high and medium quality soils support about 400 million people or about 45% of African population. Fifty five percent of the land in Africa is unsuitable for any kind of cultivated agriculture except nomadic grazing (Eswaran et al. 1996). These are largely the deserts, which include salt flats, dunes and rock lands, and the steep to very steep lands. About 30% of the population (or about 250 million) are living or dependent on these land resources.

During the past three decades, the paradigms underlying the use of fertilizers and soil fertility management research and development efforts have undergone substantial change due to experiences gained with specific approaches and changes in the overall social, economic, and political environment (Sanchez 1994). During the 1960s and 1970s, an external input paradigm was driving the research and development agenda. The appropriate use of external inputs, namely fertilizers, lime, or irrigation water, was believed to be able to alleviate any constraint to crop production. Following this paradigm together with the use of improved cereal germplasm, the 'Green Revolution' boosted agricultural production in Asia and Latin America. However, the application of the 'Green Revolution' strategy in Africa resulted in minor achievements only due to a variety of reasons (IITA 1992). One of the reasons is the abolition of the fertilizer subsidies in Africa imposed by structural adjustment programs (Smaling 1993). In the 1980s, the balance shifted from mineral inputs only to Low External Input Sustainable Agriculture (LEISA) where organic resources were believed to enable sustainable agricultural production. After a number of years of investment in research activities evaluating the potential of LEISA technologies, such as alley cropping or live-mulch systems, several constraints were identified both at the technical (e.g., lack of sufficient organic resources) and the socio-economic level (e.g., labour intensive technologies) (Vanlauwe 2004).

In this context, Sanchez (1994) revised his earlier statement by formulating the Second Paradigm for tropical soil fertility research: 'Rely more on biological processes by adapting germplasm to adverse soil conditions, enhancing soil biological activity and optimizing nutrient cycling to minimize external inputs and maximize the efficiency of their use'. This paradigm recognized the need for both mineral and organic inputs to sustain crop production, and emphasized the efficient use of all inputs. The need for organic and mineral inputs was advocated: (1) both resources fulfill different functions to maintain plant growth, (2) under most small-scale farming conditions, neither of them is available or affordable in sufficient quantities to be applied alone, and (3) several hypotheses could be formulated leading to added benefits when applying both inputs in combination. The second paradigm also highlighted the need for improved germplasm, and as in earlier days, more emphasis was put on the nutrient supply side without worrying too much about the soil demand for these nutrients. Obviously, optimal synchrony or use efficiency requires both supply and demand to function optimally.

More recently, the shift in paradigm was towards the Integrated Natural Resource Management (INRM), which combined organic and mineral inputs accompanied by a shift in approaches towards involvement of the various stakeholders in the research and development process. One of the important lessons learned was that the farmers' decision making process was not merely driven by the soil and climate but by a whole set of factors cutting across the biophysical, socio-economic, and political domain (Izac 2000). Currently, soil fertility research and strategy focus on the new paradigm of Integrated Soil Fertility Management (ISFM) which is a holistic approach in soil fertility research that embraces the full range of driving factors and consequences of soil degradation – biological, chemical, physical, social, economic, health, nutrition and political.

Agro-ecological Zones and Main Soil Types

Concepts and Definitions

Parameters used in the definition of agroecological zones (AEZ) focus on climatic and edaphic crop requirements and on the management systems under which crops are grown. Each zone has a similar combination of constraints and potentials, and serves as a focus for the targeting of recommendations designed to improve existing land-use through increasing production or by limiting land degradation.

African Agroecological Zones (AEZ)

With regard to climatic factors it is important to note that a large proportion of Africa receives less than 200 mm of rainfall per year; the Congo Basin receives over 2,000 mm of rain per year (Fig. 3.1). By differentiating areas of varying moisture conditions in Africa, Dudal (1980) identified five major agro-ecological zones on the basis of the length of growing period (LGP) (Fig. 3.2). LGP is defined in terms of the number of days when both moisture and temperature permit rainfed crop production.

The Humid Zone

This zone stretches from West to Central and East Africa where the rainfall exceeds a mean of 1,500 mm per year, temperatures ranging between an average of 24–28°C with a growing period of more than 270 days. In places, relief gives an increased rainfall. For example, Mount Cameroon rises 4,070 m above the neighbouring warm sea and receives the full force of the humid air, thus giving the highest rainfall of the continent (averaging 10,000 mm annually at Debundja).

The Sub-humid Zone

The humid to sub-humid wooded savannah zone covers areas between latitudes 5°–15° North and 5°–15° South in Central, Western and Southern Africa.

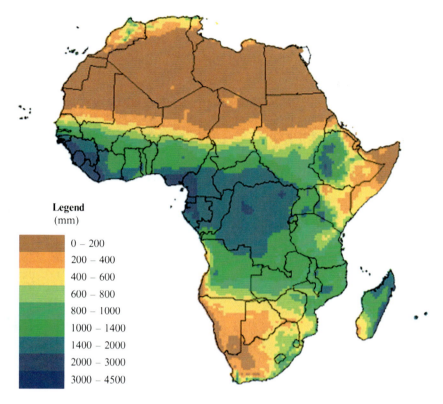

Legend
(mm)

0 – 200
200 – 400
400 – 600
600 – 800
800 – 1000
1000 – 1400
1400 – 2000
2000 – 3000
3000 – 4500

Fig. 3.1   Major rainfall zones of Africa

Areas with one or two rainy seasons of varying lengths are located within this zone. This zone has a growing period of 180–269 days.

The Semi-arid Zone

This zone covers areas between the sub-humid wooded savannah zone and the arid zone between latitudes 15°–20° north and 15°–25° south where the average rainfall ranges from 200 to 800 mm. The Sahel is part of this AEZ and has a growing period of 75–179 days.

The Arid Zone

The arid savannah zone of Africa covers extensive areas north of latitude 20°N and south of latitude 20°S where the average annual rainfall is less than 200 mm. The zone includes the vast Sahara desert and the Namibia, Kalahari and Karroo deserts. The arid zone has a growing period of <90 days.

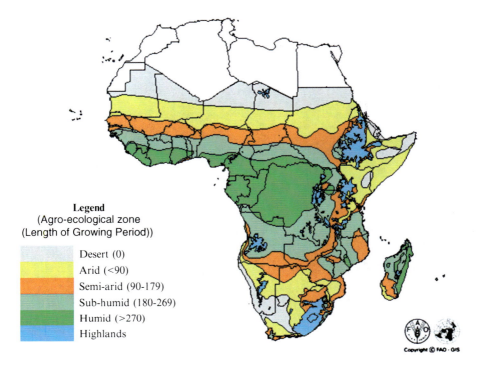

Legend
(Agro-ecological zone
(Length of Growing Period))

Desert (0)
Arid (<90)
Semi-arid (90-179)
Sub-humid (180-269)
Humid (>270)
Highlands

Copyright © FAO · GIS

Fig. 3.2 Major agro-ecological zones based on the length of the growing period

The Mediterranean Zone

This zone embraces the extreme northern and southern parts of Africa, especially the coastal areas of Algeria, Egypt, Libya, Morocco and Tunisia. Climatic conditions are quite different from tropical Africa.

Soil Types and Constraints

The soil patterns in the five major agro-ecological zones of Africa are determined by differences in age, parent material, physiography and present and past climatic conditions. There is a strong correlation between nutrient depletion, the AEZ and dominant major soils of each of the AEZ. In the humid zones dominant soils are Ferralsols and the Acrisols. Less important in this zone are Arenosols, Nitosols and Lixisols. The sub-humid zone is characterized by the dominance of Ferralsols and Lixisols and to a lesser extent Acrisols, Arenosols and Nitosols. In the semi-arid zone Lixisols have the larger share followed by sandy Arenosols and Vertisols (Deckers 1993). The next section provides a brief description of the characteristics and potential environmental problems with respect to plant nutrition of some of the

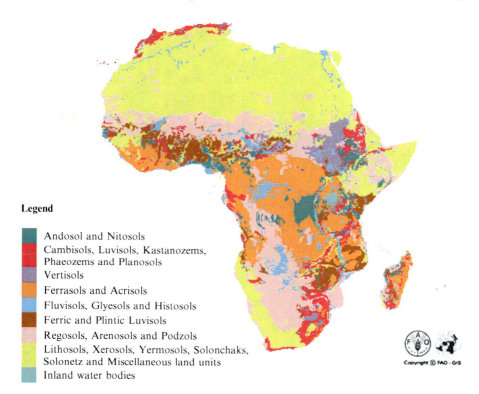

Legend

- Andosol and Nitosols
- Cambisols, Luvisols, Kastanozems, Phaeozems and Planosols
- Vertisols
- Ferrasols and Acrisols
- Fluvisols, Glyesols and Histosols
- Ferric and Plintic Luvisols
- Regosols, Arenosols and Podzols
- Lithosols, Xerosols, Yermosols, Solonchaks, Solonetz and Miscellaneous land units
- Inland water bodies

Fig. 3.3 Major soil types in Africa

dominant soils of Africa. A map showing the distribution of major soils in Africa is shown in Fig. 3.3.

Ferralsols (≈Oxisols in USDA Soil Taxonomy)

Ferralsols occupy a considerable part of Central Africa (The Democratic Republic of Congo (DRC)), Angola, Zambia, Rwanda, Burundi, Uganda, South Sudan, Central African Republic and Cameroon. In West Africa, large tracts of Ferralsols occur in Liberia and Sierra Leone (Deckers 1993). The eastern part of Madagascar is also largely characterized by Ferralsol associations. A major characteristic of Ferralsols is their advanced weathering (Deckers 1993). The capacity to supply nutrients to plants and the capacity to retain nutrients (cationic exchange capacities (CEC)) are both low. From a soil fertility point of view, this low retention capacity has marked consequences for fertilizer management. Inorganic fertilizers, especially nitrogen, should be applied in small amounts to avoid leaching as these soils occur in high rainfall regions. Phosphate fertilizers are fixed by free iron and aluminium oxides. There is need to apply high rates of P fertilizers. Other constraints of Ferralsols include: deficiency in bases (Ca, Mg, K) which requires lime application and incapability to retain bases applied as fertilizers or by other means; presence of aluminium in Ferralsols with low pHs (<5.2) – this element is toxic for many plant species and highly active in the fixation of phosphates; presence of free manganese in acid Ferralsols, likewise toxic for a number of species; deficiency of molybdenum, especially required for the growth of legumes – hazards of iron and manganese toxicity in paddy-rice. Physically Ferralsols are excellent soils, well drained and with a good structure and deep profile. Rooting depth is almost unlimited and this makes up for their relatively low water holding capacity (Van Wembeke 1974).

Acrisols (≈Ultisols in USDA Soil Taxonomy)

Acrisols occur in the southern part of the sub-humid zone of West Africa and in southern Guinea, most of Côte d'Ivoire, Southern Ghana, Togo, Benin and Nigeria and central Cameroon. In East Africa Acrisols occur in the humid parts of Tanzania (Deckers 1993). Acrisols have a high water holding capacity but the higher density of the second horizon may limit biological activity and root penetration (Deckers 1993). Although Acrisols are less weathered than Ferralsols, mineral reserves are low. Leaching is a problem in these soils and boron and manganese are often deficient. High aluminium contents may lead to phosphate fixation. The structure of the surface soil of Acrisols is weak and internal drainage may be hampered by the compact textural B horizon. Special care is therefore needed to protect Acrisols from soil erosion. The addition of lime and organic matter may be needed to ensure sustained production.

Nitosols (≈Plaeudults, Paleustults, Plaeudalfs, Paleustalfs in USDA Soil Taxonomy)

The Nitosols extent is limited in Africa; they occur in Ethiopia, Kenya, Tanzania, and east DRC and in areas of volcanic activity in the Rift Valley Zone (Deckers 1993). Nitosols have a clay-rich subsoil that is characterized by a good soil structure and they have a higher fertility level than Acrisols (Deckers 1993). The key to the high fertility of the Nitosols is the clay in the subsoil which can retain considerable amounts of plant nutrients. Phosphate fixation is common and manganese toxicity may be a problem in the more acid Nitosols. The water holding capacity of Nitosols is favourable because of the high clay content in the subsoil, and these soils have a typically open structure which allows crop roots to penetrate very deeply into the profile.

Lixisols (≈Alfisols, Oxic Kandiudalfs in USDA Soil Taxonomy)

Lixisols form a belt in West Africa between the Acrisols and the Arenosols. Other important areas of occurrence include south-east Africa and Madagascar (Deckers 1993). Lixisols, like Acrisols and Nitosols, have a clay accumulation horizon with a low capacity to store plant nutrients, but which is well-saturated with cations. The soil-pH of Lixisols is therefore medium to high and aluminium toxicity does not occur. Because of the low storage capacity for cations, Lixisols may become depleted quickly under agricultural use, though their physical characteristics are generally better than those of Acrisols.

Arenosols (≈Psamments in USDA Soil Taxonomy)

Arenosols form an almost continuous belt in West Africa, stretching from Northern Senegal, Mauritania, central Mali, and Southern Niger, through Chad to eastern Sudan. Other important Arenosol areas include parts of Botswana, Angola and South-west DRC (Deckers 1993). These soils are also dominant in the North African countries. The soil material of Arenosols is mainly composed of quartz, with a low water holding capacity, low nutrient content, low nutrient retention capacity and deficiency of minor nutrient elements including zinc, manganese, copper and iron (Deckers 1993). Deficiency of sulphur and potassium is common in Arenosols while fertilizer efficiency is hampered by severe leaching, especially of nitrogen and potassium. Arenosols tend to be weakly structured, which explains compaction of the subsoil and water/wind erosion of the topsoil. In dry areas, Arenosols contain more bases, but the poor water holding capacity places a severe limit on crop growth and performance.

Vertisols (Vertisols in USDA Soil Taxonomy)

The largest extent of Vertisols occurs in the semi-arid and sub-humid zones of Sudan and Ethiopia, and in Tanzania close to Lake Victoria (Deckers 1993). Vertisols are characterized by a high content of shrinking and swelling clays (Deckers 1993). During the rainy season, they expand and surface flooding becomes a problem. In the dry season, the clay shrinks and large deep cracks develop. Tillage is hampered by stickiness when wet and hardness when dry. A very narrow range exists between moisture stress and water excess. The permeability is low when moist, making them sensitive to erosion in the absence of vegetative cover. The physical condition of Vertisols is greatly influenced by the level of soluble salts and/or adsorbed sodium. Phosphorus availability is generally low. Due to the flooding in the rainy season, the efficiency of nitrogen fertilizer applications may be very low due to high nitrogen losses under waterlogged conditions.

Gleysols and Fluvisols (≈Aquic Suborders and Fluvents in USDA Soil Taxonomy)

Gleysols are found in equatorial Africa and inland valleys across Africa. These are soils with signs of excess wetness. The parent material is characterized by a wide range of unconsolidated materials, mainly fluvial, marine and lacustrine sediments

Table 3.1 Main soil types, extent, constraints and countries covered

Soil type	Hectare (ha)	Percentage of total land (%)	Main constraints	Main countries covered
Andosol and Nitosols	117,123,121	3.8	Fertile (volcanic ash), high P-fixation, Mn toxicity, medium water and nutrient retention	Rift valley (Ethiopia, Kenya, Tanzania, Zaire)
Cambisols, Luvisols, Kastanozems, Phaeozems and Planosols	211,348,146	6.8	Low to moderate nutrients content	Mediterranean countries
Vertisols	98,985,811	3.2	Heavy soils, medium mineral reserves, erodibility and flooding	Sudan, Ethiopia, South Africa, Lesotho
Ferrasols and Acrisols	500,910,947	16.2	Low nutrients content, Weathered, Al and Mn toxicity, high P-fixation, low nutrient and water retention, susceptible to erosion	DRC, Angola, Zambia, Rwanda, Burundi, Uganda, Sudan, Central Afr. Rep., Cameroon, Liberia, Sierra Leone and Madagascar Sub-humid zone of West Africa
Fluvisols, Glyesols and Histosols	132,037,611	4.3	Poor to moderate drainage	West, Central and southern Africa
Ferric and Plintic Luvisols	179,786,479	5.8	Low nutrients content	Western and southern Africa
Regosols, Arenosols and Podzols	579,101,963	18.7	Mainly quartz, low water and nutrient holding capacity, wind erosion. Poor soils with nutrients leaching	West Africa/Sahel, Sudan, Botswana, Angola and DRC, North Africa
Lithosols, Xerosols, Yermosols, Solonchaks, Solonetz and Miscellaneous land units	1,244,513,150	40.3	Shallow soils. Soils subject to drought Presence of salt	North African countries, South Africa, Namibia, Somalia, Sahel
Inland water bodies	27,230,091	0.9	Flood zones	
TOTAL	3,091,037,319	100		

with basic to acidic mineralogy. These soils occur mainly in depressional areas and low landscape positions with shallow groundwater. The main obstacle to use of Gleysols is the necessity to install a drainage system designed to either lower the groundwater table, or intercept seepage or surface runoff water. Drained Gleysols can be used for arable cropping, dairy farming or horticulture. Soil structure will be destroyed for a long time if wet soils are cultivated. Gleysols in (depression) areas with unsatisfactory drainage possibilities are therefore best kept under a permanent grass cover or (swamp) forest. Gleysols can be put under tree crops only after the water table has been lowered with deep drainage ditches. Gleysols in the tropics and subtropics are widely planted to rice.

Soils in the Mediterranean Region

The soils in the Mediterranean region are unique. These are Rendzinas, Phaeozems, Cambisols, Kastanozems, Arenosols and Solonchaks and they result from the combination of dry summer seasons and the prevalence of calcareous material. These soils are dominated by carbonates and are often clayey and rich in organic matter and calcium. They have good structure, well drained and have adequate available water capacity. Leached soils such as Ferralsols are rare. The main impediments of Mediterranean soils are due to excessive limestone and disproportionate soluble salts, in addition to the scarcity of water, particularly in arid and semi-arid areas. From agricultural stand point, Mediterranean soils are fertile.

Table 3.1 below summarizes the extent of the main soil types, constraints and countries covered.

Land Degradation and Soil Productivity

Land Degradation

Land degradation is defined by FAO (2002) as the loss of production capacity of land in terms of loss of soil fertility, soil bio-diversity and degradation of natural resources. Land degradation is widespread and it affects soils and landscapes functioning and human welfare (Thiombiano 2000). At least 485 million Africans are affected by land degradation, making land degradation one of the continent's urgent development issue with significant costs: Africa is burdened with a $9.3 billion annual cost of desertification. While the cumulative loss of crop productivity from land degradation worldwide between 1945 and 1990 has been estimated at 5%, as much as 6.2% of productivity has been lost in SSA. An estimated $42 billion in income and 6 million hectares of productive land are lost every year due to land degradation and declining agricultural productivity (UNDP/GEF 2004). Globally, Africa suffered a net loss of forests exceeding four million hectares per year between

Table 3.2 Global estimates of agricultural land degradation by region

	Total land (m ha)	Degraded land (m ha)	Percent degraded
Africa	187	121	65
Asia	536	206	38
S. America	142	64	45
Central America	38	28	74
North America	236	63	26
Europe	287	72	25
Oceania	49	8	16
World	1475	562	38

Source: Scherr (1999), Oldeman et al. (1992)

2000 and 2005, according to FAO. This was mainly due to conversion of forest lands to agriculture. Forest cover reduced from 656 million hectares (ha) to 635 million ha during this period.

Causes for land degradation are: human population growth, poor soil management, deforestation, insecurity in land tenure, variation of climatic conditions, and intrinsic characteristics of fragile soils in diverse agro-ecological zones. Various agricultural and non agricultural uses of soils are pointed out to have a negative impact on African lands, due to the lack of appropriate land use planning and to the mismanagement of natural resources by land users, particularly by resource poor farmers.

Land degradation is the most serious threat to food production, food security and natural resource conservation in Africa. The population is trapped in a vicious cycle between land degradation and poverty, and the lack of resources and knowledge to generate adequate income and opportunities to overcome the degradation. Scientists have reported that soil loss through erosion could be about 10 times greater than the rate of natural formation, while the rate of deforestation is 30 times higher than that of planned reforestation. Wind and water erosion are important causes of degradation but soil fertility decline (largely invisible and a gradual process) is also important.

Land degradation is caused by soil water erosion (46%), wind erosion (36%), and loss of nutrients (9%), physical deterioration (4%), and salinization (3%). Overgrazing (49%), followed by agricultural activities (24%), deforestation (14%), and overexploitation of vegetative cover (13%) constitute the primary causes of land degradation in rural areas. This degradation reduces the capacity to increase food production. Yield losses due to land degradation in Africa range from 2% decline over several decades to catastrophic, reaching up to 50% (Scherr 1999). Crop yield loss due to erosion in 1989 was 8% for Africa as a whole, which makes the fight against land degradation a fight against poverty.

Worldwide, the area of degraded soils is extensive (Table 3.2) and the effects of degradation are evident in many parts of the continent with degradation-prone soils, and unsustainable intensification in many of the densely populated zones (Scherr 1999). The situation is especially dire in Africa, where millions of people are threatened by hunger (UNDP/GEF 2004). It is estimated that since the 1950s,

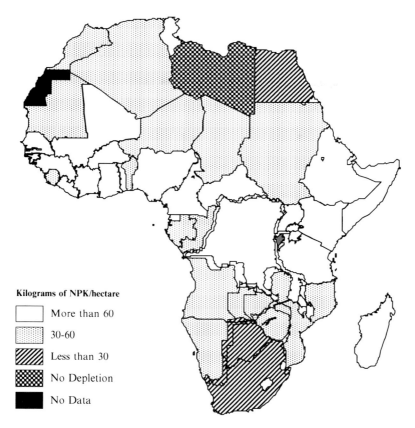

Fig. 3.4 Average annual nutrient depletion (NPK) in Africa between (1993–1995) (*Source*: Henao and Baanante 1999b)

Africa has lost about 20% of its soil productivity irreversibly due to degradation (Dregne 1990).

Soil-fertility depletion in smallholder farms is a fundamental biophysical root cause of the declining per capita food production; it has largely contributed to poverty and food insecurity. Over 132 million tons of N, 15 million tons of P and 90 million tons of K have been lost from cultivated land in 37 African countries in 30 years (Smaling 1993). Nutrient loss is estimated to be 4.4 million tons N, 0.5 million tons P and 3 million tons K every year from the cultivated land (Sanchez et al. 1997). These rates are several times higher than Africa's annual fertilizer consumption (excluding South Africa) of 0.8 million tons N, 0.26 million tons P, and 0.2 million tons K. The loss is equivalent to 1,400 kg ha^{-1} urea, 375 kg ha^{-1} Triple Super-phosphate (TSP) and 896 kg ha^{-1} KCl during the last three decades. Figure 3.4 shows the annual average nutrient depletion in Africa measured in kilograms per hectare per year between 1993 and 1995.

Soil organic carbon is a depletable natural resource capital and like the negative nutrient balances its decline threatens soil productivity. The concentration of organic carbon in the top soil is reported to average 12 mg kg^{-1} for the humid zone, 7 mg kg^{-1} for the sub-humid zone and 4 mg kg^{-1} or less in the semi-arid zone (Windmeijer and Andriesse 1993). The inherently low soil organic carbon is due to the low root growth of crops and natural vegetation but also the rapid turnover rates of organic materials with high soil temperature and microfauna, particularly termites (Bationo et al. 2003). There is much evidence for rapid decline of soil organic C levels with continuous cultivation of crops in Africa (Bationo et al. 1995). For the sandy soils, average annual losses in soil organic C may be as high as 5%, whereas for sandy loam soils, reported losses seem much lower, with an average of 2% (Pieri 1989). Results from long-term soil fertility trials indicate that losses of up to 0.69 t carbon hectare^{-1} year^{-1} in the soil surface layers is common in Africa even with high levels of organic inputs (Nandwa 2003).

Soil Productivity

Soil Moisture

Soil moisture stress is perhaps the overriding constraint to food production in much of Africa. Only about 14% of Africa is relatively free of moisture stress. The incidence of drought since 1975 has increased nearly fourfold (UNDP/GEF 2004). Moisture stress is not only a function of the low and erratic precipitation but also of the ability of the soil to hold and release moisture. About 10% of the soils in Africa have high to very high available water holding capacities (AWHC). These are mainly Vertisols, and other clayey soils. The 29% of soils with medium AWHC are mainly Lixisols and Ferralsols and some loamy Inceptisols and Entisols. The low AWHC class soils are the Ferralsols and other sandy loam soils. Despite their clayey textures, Ferralsols have low AWHC. The very low AWHC class soils are sandy soils such as Arenosols and other sandy and sandy loam soils. The development of conservation agriculture technologies with soil permanent cover will be of importance for the increasing and conservation of soil moisture as it has been shown in various FAO projects.

Productivity Zones

Soil properties, including soil climate, provide some preliminary information to address soil quality and to classify land according to its potential for productivity. Land can be classified as prime, high, medium, low potential lands and the unsuitable class of lands. Figure 3.5 gives the agricultural potential of African soils. Prime land comprises those soils with deep, permeable layers, with an adequate supply of nutrients, and generally do not have significant periods of moisture stress. The soils

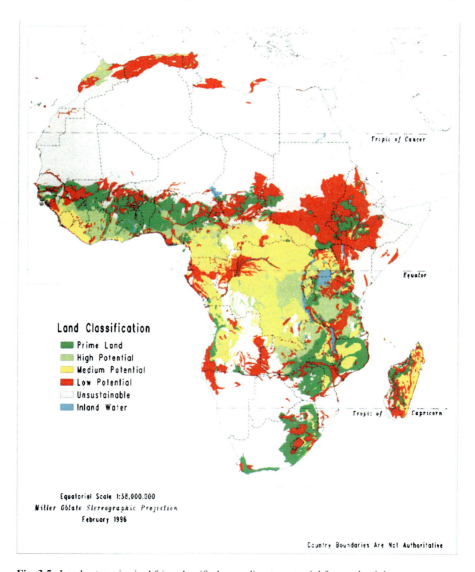

Fig. 3.5 Land categories in Africa classified according to potential for productivity

are deep, without impermeable layers, textures are loamy to clayey with good tilth characteristics, and the land is generally level to gently undulating. They occupy about 9.6% of Africa and they occupy significant areas in West Africa south of the Sahel, in East Africa mainly in Tanzania, and in Southern African countries of Zambia, Zimbabwe, South Africa, and Mozambique.

High potential soils are similar to the prime soils but have some minor limitations such as extended period of moisture stress, sandy or gravelly materials, or with root restricting layers in the soil. The high potential lands occupy an area of about 6.7%.

Medium and low potential lands, which occupy 28.3% of the surface, have Key constraints for low-input agriculture. These lands have a major soil constraint and one or more minor constraints which can be corrected. Constraints include adverse soil physical properties including surface soil crusting, impermeable layers, soil acidity and specifically subsoil acidity, salinity and alkalinity, and high risks of wind and water erosion. The large contiguous areas of Central and West Africa are considered as medium potential, due to the presence of acid soils and soils which fix high amounts of P. With an inherently low soil quality, low-input agriculture can be equated to potential soil degradation. These are some of the priority areas for technical assistance and the implementation of appropriate soil management technologies.

The unsustainable class of lands are those which are considered to be fragile, easily degraded through bad management, and in general are not productive or do not respond well to management. These occupy about 55% of the African continent. They are generally erodible and require high investments.

Profitability of the Use of Fertilizers in African Agro-ecosystems

There is ample evidence that increased use of inorganic fertilizers has been responsible for an important share of world-wide agricultural productivity growth. Fertilizer was as important as seed in the Green Revolution contributing as much as 50% of the yield growth in Asia (Hopper 1993). Several studies have found that one third of the cereal production worldwide is due to the use of fertilizer and related factors of production (Bumb 1995, citing FAO). Van Keulen and Breman (1990) and Breman (1990) stated that the only real cure against land hunger in the West African Sahel lay in increased productivity of the arable land through the use of external inputs, mainly inorganic fertilizers. Pieri (1989) reporting on fertilizer research conducted from 1960 to 1985, confirmed that inorganic fertilizers in combination with other intensification practices, had tripled cotton yields in West Africa from 310 to 970 kg ha^{-1}. There are numerous cases of strong fertilizer response for maize in East and Southern Africa (Byerlee and Eicher 1997).

The data in Table 3.3 summarizes multi-site response to soil fertility improvement and clearly demonstrates the importance of fertilizers in maize yield improvement in different AEZ and soil types in Africa. Maize yield increase over the control due to NPK fertilizer application from six AEZ and averaged over 4 years was 149% but when the soil was amended with lime and manure yield response over the control increased to 184% (Mokwunye et al. 1996). Similarly higher yield improvements have been observed in East (Qureish 1987) and Southern (Mtambanengwe and Mapfumo 2005) African countries.

Despite the above response to fertilizer application, studies have shown that there are great on-farm soil fertility gradients and the yields are bound to vary greatly even on the same production unit. Prudencio (1993) observed such fertility gradients between the fields closest to the homestead (home gardens/infields) and

Table 3.3 Maize response to organic and inorganic fertilizer application in selected sites in East, West and Southern Africa

Site	Treatment	Yield (t/ha)	% Yield increase
[a]West Africa (Multi sites 3–6 years average)	Control	1.51	–
	TSP + N + K	3.172	110
	N + K	2.319	54
	P + K	2.426	61
	P + N + K	3.765	149
	P + N + K + lime (500 kg ha[-1] every 3 years)	3.794	151
	Crop residue (CR)	1.999	32
	Manure (10 t ha[-1] every 3 years)	2.497	65
	P + N + K + Mg + Zn	3.880	157
	P + N + K + Mg + Zn + Lime	4.006	165
	P + N + K + Mg + Zn + CR	4.083	170
	P + N + K + Mg + Zn + Manure + Lime	4.289	184
[b]East Africa (1981–1985)	Control	1.9	–
	Crop residues	2.5	32
	Manure (5 t)	3.5	84
	Fertilizer (60 kg N, 25 kg P)	4.1	116
	Manure (10 t)	4	111
	Fertilizer (120 kg N, 50 kg P)	4.6	142
	Manure (5 t + Fertilizer 60 kg N, 25 kg P)	5.2	174
[c]Southern Africa	Control	0.729	–
	N + P + K	2.194	201
	Termitaria + N + P	2.229	206
	Cattle manure + N + P	2.644	262
	Maize stover + N + P	1.575	116
	Fresh litter + N + P	2.553	250
	Crotalaria juncea + N + P	2.496	242

[a]Mokwunye et al. (1996) – Results from six different AEZ in Togo
[b]Qureish (1987) – Results from Kabete, Kenya
[c]Mtambanengwe and Mapfumo (2005) – Results from Chinyika Zimbabwe

those furthest (bush fields/outfields). Soil organic carbon contents of between 11 and 22 g kg[-1] have been observed in home gardens compared to 2–5 g kg[-1] soil in the bush fields. Similarly, higher total organic nitrogen, available phosphorus and exchangeable potassium have also been observed in the home gardens compared to the bush fields. Fofana et al. (2008), in a study in West Africa, observed that grain yields across years and fertilizer treatment averaged 0.8 t ha[-1] on outfields and 1.36 t ha[-1] on infields (home gardens). Recovery of fertilizer N (RFN) applied varied considerably among the treatments and ranged from 17% to 23% on outfields and 34–37% on infields. Similarly, average recovery of fertilizer P applied (RFP) across treatments was 31% in the infields compared to 18% in the outfields over the 3-year mono cropping period. These results indicate higher inherent soil fertility and nutrient use efficiency in the infields compared to the outfields and underlines the importance of soil organic carbon in improving fertilizer use efficiency. Once soils are degraded and poor in organic matter, the response to fertilizer is less and the recovery of applied fertilizers is very low.

Despite the fact that deficiency of P is acute on the soils of Africa, local farmers use very low P fertilizers because of high cost and problems with availability. The use of locally available phosphate rock (PR) could be an alternative to imported P fertilizers. For example, Bationo et al. (1987) showed that direct application of local PR may be more economical than imported water-soluble P fertilizers. Bationo et al. (1990) showed that Tahoua PR from Niger is suitable for direct application, but Parc-W from Burkina Faso has less potential for direct application. The effectiveness of local phosphate rock depends on its chemical and mineralogical composition (Chien and Hammond 1978). Phosphate rocks can be used as a soil amendment and the use of water soluble P can be more profitable.

Phosphorus placement can drastically increase P use efficiency as shown with pearl millet and cowpea in an experiment involving broadcast (BC) and/or hill placed (HP) of different P sources. For pearl millet grain P use efficiency for broadcasting SSP at 13 kg P ha^{-1} was 23 kg kg^{-1} but hill placement of SSP at 4 kg P ha^{-1} gave a PUE of 83 kg kg^{-1} P. The PUE of 15-15-15 broadcast was 29 kg grain kg^{-1} P, whereas the value increased to 71 kg kg^{-1} P when additional SSP was applied as hill placed at 4 kg P ha^{-1} and 102 when only HP of 4 kg P ha^{-1} of 15-15-15 was used. Hill placement of small quantities (4 kg ha^{-1}) of P attains the highest use efficiency with the efficiency decreasing with increasing quantity of P.

Farm-level fertilizer prices in Africa are among the highest in the world. The cost of 1 metric ton of urea, for example costs about US$90 in Europe, US$500 in Western Kenya and US$700 in Malawi. These high prices can be attributed to the removal of subsidies, transaction costs, poor infrastructure, poor market development, inadequate access to foreign exchange and credit facilities and lack of training to promote and utilize fertilizers. For example it costs about $15, $30 and $100 to move 1 t fertilizer 1,000 km in the USA, India and SSA respectively.

Contrary to conventional wisdom, there are examples of evidence of fertilizer response and profitability in Africa that compare favourably to those in other parts of the world. Yanngen et al. (1998) in a comprehensive study of fertilizer profitability in Africa found out that among the cereal crops covered, maize and irrigated rice exhibited the strongest incentives to fertilizer application. The yield response and the profit incentives for rice and maize have been observed to be equal or higher than what was obtained in Latin America and Asia respectively (Yanggen et al. 1998) (Table 3.4). These results and other numerous responses from site specific studies are clear evidence that the use of inorganic fertilizer could be in deed a profitable investment.

Table 3.4 Fertilizer incentives: Summary of key indicators by crop and region

Type of crop	Region	Kg of output/kg of nutrient use (efficiency)			Profit incentives (V/C ratio)	
		Typical	Min	Max	Min	Max
Maize	East and Southern Africa	17	2	52	1	15
	West Africa	15	0	54	0.69	26
	Latin America	10	5	18	1.2	5.3
Cotton	East and Southern Africa	5.8	0	7	0.00	3.1
	West Africa	5	2	12	0.61	3.7
Rice (irrigated)	West Africa	12	7	16	1.6	3.97
	Asia	11	7.7	33.6	1.5	3.1
Sorghum	East and Southern Africa	10	4	21	1.5	2.6
	West Africa	7	3	14	1	18
	Latin America	7	2.8	21		

Adapted from Yanngen et al. (1998)

A major problem to effective utilization of fertilizers has been 'pan-territorial/blanket' recommendations that fail to take into account differences in resource endowment (soil type, labour capacity, climate risk, etc.). The situation is exacerbated by the failure to revise recommendations following dramatic changes in the input/output price ratios due to subsidy removal and devaluation of currencies. Farmers using inorganic fertilizers experiment with different rates and methods of application. In West Africa, for example, farmers have adopted the 'microdose' technology that involves strategic application of small doses of fertilizer 4 kg P ha^{-1} and seed (Tabo et al. 2006). This rate of fertilizer application is only one-third of the recommended rates for the areas. In all the project study sites in the three West African countries (Burkina Faso, Mali and Niger), grain yields of millet and sorghum increases were up to 43–120% when using fertilizer "micro-dosing". The incomes of farmers using fertilizer "micro-dosing" and inventory credit system or "Warrantage" increased by 52–134%. Small amounts are more affordable for farmers, give an economically optimum (though not biologically maximum) response, and if placed in the root zone of these widely-spaced crops rather than uniformly distributed, result in more efficient uptake (Bationo and Buerkert 2001). The successful experience has shown that adoption of micro-dose technology requires supportive and complementary institutional innovation and market linkage. Organized farmer groups provide access to post-harvest credit provided on the basis of storage of grain as collateral ("warrantage"), enabling farmers to sell crops later in the season for higher prices and higher profits. They also provide greater access to fertilizer, leading to higher sustained yields, and income.

Variability of rainfall is a critical factor in efficiency of fertilizers and in determining risk-aversion strategies of farmers in Africa. The tendency of African rainfall to be both spatially and temporally concentrated has important implications for fertilizer use. A survey of available data found Africa levels of available water from rainfall at 12.7 cm year^{-1} compared to North America at 25.8, South America at

Table 3.5 Water use (*WU*), grain yield (*Y*) and water use efficiency (*WUE*) for millet at Sadore and Dasso (Niger)

Treatment	Sadore			Dosso		
	WU (mm)	Y (kg ha^{-1})	WUE	WU (mm)	Y (kg ha^{-1})	WUE
Fertilizer	382	1570	4.14	400	1,700	4.25
Without fertilizer	373	460	1.24	381	780	2.04

ICRISAT (1985)

Table 3.6 Effect of 'Zai' on sorghum yields

Technology	Sorghum yield (kg ha^{-1})	Yield increase (%)
Only planting pits (Zai)	200	–
Zai + Cattle manure	700	250
Zai + Mineral fertilizers	1,400	600
Zai + Cattle manure and fertilizers	1,700	750

Reij et al. (1996)

64.8 and the world average at 24.9 (Brady 1990). Fertilizer is commonly thought to increase risk in dryland farming, but in some situations it may be risk-neutral or even risk-reducing. Phosphorus and shorter-duration millet varieties in Niger, for example cause crops to grow hardier and mature earlier, reducing damage from and exposure to drought (ICRISAT 1985–1988; Shapiro and Sanders 1998). A key constraint though is the availability of fertilizer and the incentive for adopting fertility-enhancing crop rotations in these zones (Thomas et al. 2004).

In the dry land of the Sahel, several scientists have reported that where the rainfall is more than 300 mm, the most limiting factors to crop production is nutrient and not water. At Sadore where the annual average rainfall is 560 mm, the non use of fertilizers resulted in a harvest of 1.24 kg of pearl millet grain per mm of water but the use of fertilizers resulted in the harvest of 4.14 kg of millet grain per mm of water (Table 3.5)

Many development projects have invested billions of dollars in soil and water conservation. It mostly did not include soil fertility improvement and the water harvested in this manner is not fulfilling its full potential for productivity improvement. It is well known that fertilizers are a key to improved water use efficiency as water harvesting can also improve the fertilizer use efficiency. The Zai system is widely used in West Africa for water harvesting and soil conservation. The data in Table 3.6 indicates that the use of Zai alone will not improve much the productivity (only 200 kg ha^{-1} of sorghum grain) but when the Zai is associated with manure and fertilizer large crop yield increases can be obtained (1,700 kg ha^{-1} of sorghum grain) (Table 3.6).

Conclusions

Africa has an extremely wide range of soils and climatic conditions. The soils range from stony shallow ones with meager life-sustaining capabilities to deeply weathered profiles that recycle and support large biomass. In many parts of Africa, inappropriate land use, poor management and lack of inputs have led to a decline in productivity, soil erosion, salinization and loss of vegetation. African soils are at risk, they are commonly undergoing degradation since the traditional methods used by farmers (shifting cultivation, nomadic grazing) cannot cope with the increasing needs of the ever-expanding human and livestock populations. Conservation action to halt and reverse degradation needs to be planned in detail for each land type and socio-economic circumstance. Positive developments also occur, but so far seem to be drops in an ocean of land degradation.

The very low use of inorganic fertilizer has greater negative environmental consequences than excess use of inorganic fertilizers. Organic sources are not sufficient to replace nutrients lost or removed from the soils. Increased inorganic fertilizer use would benefit the environment by reducing the pressure to convert forests and other fragile lands to agricultural uses and, by increasing biomass production, and help increasing the soil organic matter content. This organic material supplies and helps retain soil nutrients.

The present farming systems are unsustainable. Intensification is needed to feed growing populations but it must be done in a way that uses soil nutrient and water resources efficiently, and to relieve pressure on forests and other fragile lands. Technologies need to be adapted to the specific bio-physical and socio-economics circumstances of the small scale farmers in Africa. For efficient nutrient utilization, inorganic fertilizer must be combined with organic matter, water harvesting, conservation agriculture and controlling soil erosion in site-specific integrated soil fertility management strategies. These complementary activities help insure that maximum benefits are derived from each component practice.

Although significant progress has been made in research in developing methodologies and technologies for combating soil fertility depletion, the low adoption is a reason for the large difference between farmers' yields and potential yield. The costs of fertilizers can be 2–4 times higher in Africa than in the developed countries. Without a set of conducive policies, it will be hard to raise fertilizer use levels outside traditional user countries such as Egypt and South Africa. There is need to focus more on increasing the fertilizers use efficiency in order to make them more profitable. Another way to get fertilizer price affordable by the small scale farmers is the development of the local fertilizer sector. There is need for feasibility studies for the development of the fertilizer sector mainly using indigenous phosphate rocks. There is need to develop tools for the scaling up of success stories such as the use of zai technique, microdosing, small-scale water harvesting, tree regrowth for soil mulch, and others.

Macro-policy changes were executed without much understanding of the likely consequences at micro-level and hidden effect on continued erosion of the natural resource base. Structural adjustment policies resulted in the reduction of the use of external inputs, extensification of agriculture, further encroachment on land under natural vegetation, and the reduction of farmers' potential to invest in soil fertility restoration. Future research needs to look for ways and means to reverse these trends.

Interest in the quality and health of soil has grown with the recognition that soil is vital not only for production of food and fibre, but also for other ecosystem services than those directly providing tangible benefits for human welfare. Research on inorganic fertilizer should consider its effect on aspects of climate change such as green house gas (GHG) emissions and carbon sequestration, water quality, and interaction with pest and diseases. The challenge is to establish and quantify the global benefits resulting from sustainable land management. The use of decision support tools can help predict and unravel some of these challenges facing integrated sustainable management of natural resources.

References

Bationo A, Buerkert A (2001) Soil organic carbon management for sustainable land use in Sudano Sahelian West Africa. Nutr Cycl Agroecosys 61:131–142

Bationo A, Christianson CB, Mokwunye AU (1987) Soil fertility management of the millet-producing sandy soils of Sahelian West Africa: the Niger experience. Paper presented at the workshop on soil and crop management systems for rainfed agriculture in the Sudano-Sahelian zone, International Crops Research Institute for the Semi-Arid Tropics (ICRISAT), Niamey, Niger

Bationo A, Chien SH Christianson CB, Henao J, Mokwunye AU (1990) A three year evaluation of two unacidulated and partially acidulated phosphate rocks indigenous to Niger. Soil Sci Soc Am J 54:1772–1777

Bationo A, Buerkert A, Sedogo MP, Christianson BC, Mokwunye AU (1995) A critical review of crop residue use as soil amendment in the West African Semi-arid tropics. In: Powell JM, Fernandez-Rivera S, Williams TO, Renard C (eds) Livestock and sustainable nutrient cycling in mixed farming systems of Sub-Saharan Africa, vol 2. International Livestock Centre for Africa (ILCA), Addis Ababam, pp 305–322. Technical papers. In: Proceedings of an international conference, 22–26 Nov 1993, Addis Ababa

Bationo A, Mokwunye U, Vlek PLG, Koala S, Shapiro BI (2003) Soil fertility management for sustainable land use in the West African Sudano-Sahelian zone. In: Gichuru MP et al (eds) Soil fertility management in Africa: a regional perspective. Academy of Science Publishers (ASP) and Tropical Soil Biology and Fertility of CIAT, Nairobi

Brady N (1990) The nature and properties of soils, 10th edn. MacMillan, New York

Breman H (1990) No sustainability without external inputs. Sub-Saharan Africa; beyond adjustment. Africa Seminar. Ministry of foreign affairs, DGIS, The Hague, pp 124–134

Bumb B (1995) Global fertilizer perspective, 1980–2000: the challenges in structural transformations. Technical Bulletin T-42. International Fertilizer Development Center, Muscle Shoals

Byerlee D, Eicher C (eds) (1997) Africa's emerging maize revolution. Lynn Rienner, Boulder

Chien SH, Hammond LL (1978) A simple chemical method for evaluating the agronomic potential of granulated phosphate rock. Soil Sci Soc Am J 42:615–617

Deckers J (1993) Soil fertility and environmental problems in different ecological zones of the developing countries in sub Saharan Africa. In: Van Reuler H, Prins WH (eds) The role of plant nutrients and sustainable food production in sub-Saharan Africa. Vereniging van Kunstmest Producenten, Laidschendam

Dregne HE (1990) Erosion and soil productivity in Africa. J Soil Water Conserv 45:431–436

Dregne HE, Chou NT (1994) Global desertification dimensions and costs. In: Dregne HE (ed) Lubbock degradation and restoration of arid lands. Texas Technical University, Lubbock

Dudal R (1980) Soil related constraints to agricultural development in the tropics. In: Proceedings symposium on properties for alleviating soil related constraints to food production, IRRI-Cornell University, Los Banos, 1980, pp 23–40

Eswaran H, Almaraz R, van den Berg E, Reich P (1996) An assessment of the soil resources of Africa in relation to productivity. World Soil Resources, Soil Survey Division, USDA Natural Resources Conservation Service, Washington, DC, Received 28 Feb 1996

FAO (2002) Land degradation assessment in drylands. FAQ, Rome, 18 pp

FAO (Food and Agriculture Organization of the United Nations) (1978) Report on the agro-ecological zones project – vol 1: methodology and results for Africa. FAO, Rome, 158 pp. World Soil Resources Project 48

Fofana B, Wopereis MCS, Bationo A, Bremen H, Mando A (2008) Millet nutrient use efficiency as affected by natural soil fertility, mineral fertilizer use and rainfall in the west African Sahel. Nutrient Cycling in Agroecosystems 61:25–36

Henao J, Baanante C (1999a) Nutrient depletion in the agricultural soils of Africa. IFPRI 2020 Vision. Brief No. 62

Henao J, Baanante C (1999b) Estimating rates of nutrient depletion in soils of agricultural lands in Africa. International Fertilizer Development Centre, Muscle Shoals

Hopper WD (1993) Indian agriculture and fertilizer: an outsiders observations: keynote address to the FAI Seminar on emerging scenario in fertilizer and agriculture: global dimensions. FAI, New Delhi

ICRISAT (1985–1988) ICRISAT Sahelian Center annual report, 1984. International Crops Research Institute for the Semi-Arid Tropics, Patancheru

ICRISAT (International Crop Research Institute for the Semi-Arid Tropics) (1985) Annual report 1984. ICRISAT Sahelian Center, Niamey

IITA (International Institute of Tropical Agriculture) (1992) Sustainable food production in sub-Saharan Africa: 1. IITA's contributions. IITA, Ibadan

Izac A-MN (2000) What paradigm for linking poverty alleviation to natural resources management? In: Proceedings of an international workshop on integrated natural resource management in the CGIAR: approaches and lessons, Penang, 21–25 Aug 2000

Mokwunye AU, de Jager A, Smaling EMA (1996) Restoring and maintaining the productivity of West African soils: key to sustainable development. IFDC-Africa, LEI-DLO and SC-DLO, International Fertilizer Development Centre, Lome´, Togo, pp 94

Mtambanengwe F, Mapfumo P (2005) Effects of organic resource quality on soil profile N dynamics and maize yields on sandy soils in Zimbabwe. Plant Soil 281(1–2):173–190

Nandwa SM (2003) Perspectives on soil fertility in Africa. In: Gichuru MP et al (eds) Soil fertility management in Africa: a regional perspective. Academy of Science Publishers (ASP) and Tropical Soil Biology and Fertility of CIAT, Nairobi

Oldeman LR, Hakkeling RTA, Sombroek WG (1992) World Map of the status of human-induced soil degradation: an explanatory note. ISRIC, Wageningen

Pieri C (1989) Fertilité des terres de savane. Bilan de trente ans de recherche et de développement agricoles au Sud du Sahara. Ministère de la Coopération. CIRAD, Paris, 444 pp

Prudencio CY (1993) Ring management of soils and crops in the West African semi-arid tropics: the case of the mossi farming system in Burkina Faso. Agr Ecosyst Environ 47:237–264

Qureish JN (1987) The cumulative effects of N-P fertilizers, manure and crop residues on maize grain yields, leaf nutrient contents and some soil chemical properties at Kabete. Paper presented at National Maize Agronomy Workshop, CIMMYT, Nairobi, Kenya, 17–19 Feb 1987

Reij C, Scoones I, Toulmin C (1996) Sustaining the soil: indigenous soil and water conservation in Africa. Earthscan, London

Sanchez PA (1994) Tropical soil fertility research: towards the second paradigm. In: Inaugural and state of the art conferences. Transactions 15th world congress of soil science, Acapulco, pp 65–88

Sanchez PA, Shepherd KD, Soule MJ, Place FM, Buresh RJ, Izac AMN, Mokwunye AU, Kwesiga FR, Ndiritu CG, Woomer PL (1997) Soil fertility replenishment in Africa: an investment in natural resource capital. In: Buresh RJ, Sanchez PA, Calhoun F (eds) Replenishing soil fertility in Africa, vol 51, SSSA special publication. Soil Science Society of America, Madison, pp 1–46

Scherr SJ (1999) Past and present effects of soil degradation. In: Scherr SJ (ed) Soil degradation – a threat to developing-country food security by 2020, International Food Policy Research Institute 2020 Discussion Paper 27, Washington DC, pp 13–30

Shapiro BI, Sanders JH (1998) Fertilizer use in semiarid West Africa: profitability and supporting policy. Agr Syst 56:467–482

Smaling EMA (1993) An agro-ecological framework of integrated nutrient management with special reference to Kenya. Wageningen Agricultural University, Wageningen

Smaling EMA (1995) The balance may look fine when there is nothing you can mine: nutrient stocks and flows in West African soils. In: Gerner H, Mokwunye AU (eds) Use of phosphate rock for sustainable agriculture in West Africa. Proceedings of a seminar on the use of local mineral resources for sustainable agriculture in West Africa, held at IFDC-Africa, 21–23 Nov 1994

Tabo R, Bationo A, Gerard B, Ndjeunga J, Marchal D, Amadou B, Annou MG, Sogodogo D, Taonda J-BSibiry, Hassane O, Diallo MK, Koala S (2006) Improving cereal productivity and farmers' income using a strategic application of fertilizers in West Africa. In: Bationo A, Waswa BS, Kihara J, Kimetu J (eds) Advances in integrated soil fertility management in Sub Saharan Africa: challenges and opportunities. Springer, Dordrecht

Thiombiano L (2000) Etude des facteurs édaphiques et pédopaysagiques dans le développement de la désertification en zone sahélienne du Burkina Faso. Thèse de Doctorat d'Etat. Université de Cocody. Volume I, 200p + Annexes

Thomas RJ, El-Dessougi H, Tubeileh A (2004) Soil fertility and management under arid and semi-arid conditions. In: Uphoff N (ed) Biological approaches for sustainable soil systems. Marcel Dekker, New York (in press)

UNDP/GEF (2004) Reclaiming the land sustaining livelihoods: lessons for the future. United Nations Development Fund/Global Environmental Facility, Nov 2004

Van Keulen H, Breman H (1990) Agricultural development in the West African Sahelian region: a cure against land hunger? Agr Ecosyst Environ 32:177–197

Van Wembeke A (1974) Management properties of ferralsols, vol 23, Soils bulletin. FAO, Rome

Vanlauwe B (2004) Integrated soil fertility management research at TSBF: the framework, the principles, and their application. In: Bationo A (ed) Managing nutrient cycles to sustain soil fertility in sub-Saharan Africa. Academy Science Publishers, Nairobi

Wallace MB, Knausenberger WI (1997) Inorganic fertilizer use in Africa: environmental and economic dimensions environmental and natural resources policy and training (EPAT) Project applied research, technical assistance and training. Winrock International Environmental Alliance, Arlington, Sept 1997

Windmeijer PN, Andriesse W (1993) Inland valleys in West Africa: an agroecological characterization of rice-growing environments. Publication 52. ILRI, International Institute for Land Reclamation and Improvement, Wageningen

Yanngen D, Kelly V, Reardon T, Naseem A (1998) Incentives for fertilizer use in sub-Saharan Africa: a review of empirical evidence on fertilizer response and profitability. International Development Working Paper No 70. Michigan State University, East Lansing

Chapter 4
Uncertainties in Simulating Crop Performance in Degraded Soils and Low Input Production Systems

James W. Jones, J. Naab, Dougbedji Fatondji, K. Dzotsi, S. Adiku, and J. He

Abstract Many factors interact to determine crop production. Cropping systems have evolved or been developed to achieve high yields, relying on practices that eliminate or minimize yield reducing factors. However, this is not entirely the case in many developing countries where subsistence farming is common. The soils in these countries are mainly coarse-textured, have low water holding capacity, and are low in fertility or fertility declines rapidly with time. Apart from poor soils, there is considerable annual variability in climate, and weeds, insects and diseases may damage the crop considerably. In such conditions, the gap between actual and potential yield is very large. These complexities make it difficult to use cropping system models, due not only to the many inputs needed for factors that may interact to reduce yield, but also to the uncertainty in measuring or estimating those inputs. To determine which input uncertainties (weather, crop or soil) dominate model output, we conducted a global sensitivity analysis using the DSSAT cropping system model in three contrasting production situations, varying in environments and management conditions from irrigated high nutrient inputs (Florida, USA) to rainfed crops with manure application (Damari, Niger) or with no nutrient inputs (Wa, Ghana). Sensitivities to uncertainties in cultivar parameters accounted for about 90% of yield variability under the intensive

J.W. Jones (✉) • K. Dzotsi • S. Adiku
Agricultural and Biological Engineering Department, University of Florida,
Museum Road, PO Box 110570, Gainesville, Florida 32611, USA
e-mail: jimj@ufl.edu

J. Naab
Farming Systems Research, Savannah Agricultural Research Institute, PO Box 494,
Wa, Upper West Region, Wa, Ghana

D. Fatondji
Agro-Ecosystem, ICRISAT Sahelian Center, PO Box 12404, Niamey, Niger

J. He
College of Water Resources and Architectural Engineering, Northwest A&F University,
Yangling, Shaanxi 712100, P.R. China

J. Kihara et al. (eds.), *Improving Soil Fertility Recommendations in Africa using the Decision Support System for Agrotechnology Transfer (DSSAT)*,
DOI 10.1007/978-94-007-2960-5_4, © Springer Science+Business Media Dordrecht 2012

management system in Florida, whereas soil water and nutrient parameters dominated uncertainties in simulated yields in Niger and Ghana, respectively. Results showed that yield sensitivities to soil parameters dominated those for cultivar parameters in degraded soils and low input cropping systems. These results provide strong evidence that cropping system models can be used for studying crop performance under a wide range of conditions. But our results also show that the use of models under low-input, degraded soil conditions requires accurate determination of soil parameters for reliable yield predictions.

Keywords Crop model • Parameters • Uncertainty • Global sensitivity analysis • Water • Cultivar • Nitrogen

Introduction

Models are increasingly being used as research tools to predict outcomes of cropping systems under different climate, soil, and management conditions in both developed and developing countries. Many papers have been published on research, demonstrating that cropping system models perform adequately for the intended purposes, such as to study impacts of different cultivars, irrigation, fertility, and cultural management practices on yield and other predicted outputs. Many of these studies have emphasized the importance of incorporating climate uncertainty to adequately consider risks to production and profitability (e.g., Hammer and Muchow 1991; Thornton and Wilkens 1998). Impacts of, and adaptation to, climate change have made extensive use of crop models, and now these models are being used for simulating years of crop rotation for projecting long term changes in soil carbon and other properties that affect sustainability of production in different environments.

Typically in these studies, researchers are interested in only a few factors that may limit growth and yield, such as water and nitrogen in addition to climate. However, there may be many factors that limit production in farmers' fields and that present challenges to model users. This is particularly true in developing countries where (1) soils are low in fertility and hold very little water, (2) where farmers typically do not apply fertilizer or irrigate, (3) there is considerable annual variability in climate, (4) weeds, insects and diseases may cause considerable damage the crop, and (5) farming practices are mainly subsistence with low input. Mathews and Stephens (2002) pointed out the difficulties of obtaining inputs to operate cropping system models in developing countries. This presents one of the challenges in reliable use of cropping system models in those countries. However, there is another major challenge that has been ignored in most previous studies, even if inputs for model studies were collected – uncertainty in environmental parameters and inputs. Model developers routinely emphasize the importance of obtaining accurate cultivar coefficients in order to apply cropping system models in local studies, which suggests that without reliable values for these parameters, the models will not adequately simulate the responses to climate and management that users are studying. Little attention has been given to uncertainty analysis of other factors, such as different soil and management inputs relative to prediction of cropping system performance. Based on studies conducted in West Africa, we hypothesized that uncertainty in soil parameters, initial conditions, and nutrient inputs contribute more to prediction uncertainty than cultivar parameters in low input, rainfed cropping systems on soils with low fertility and small water holding capacities. In this study, we used the DSSAT Cropping System Model (CSM) to simulate soil processes and crop growth responses to the harsh conditions in sites with maize in Ghana (using the CERES-Maize) and for millet in Niger (using the CERES-Millet component). Results from these two sites are compared with those for maize production in a high input production system in the USA. In this study, all other management factors, such as plant density, row spacing, etc., were input as fixed values and there was minimal damage due to pests in each of the experiments.

The objectives of this study were (1) to determine sensitivities of the DSSAT – CSM model to uncertainties in parameters for three contrasting cropping systems in low-fertility, sandy soils, and (2) to estimate the uncertainty of simulated crop production as affected by uncertainties in important soil parameters, soil initial conditions, cultivar parameters, and nutrient inputs.

Methods

Experiments

Data from a total of three experiments, two of which were conducted in West Africa (Wa, Ghana and Damari, Niger) and the other in the USA (Gainesville, Florida) were selected for this study. Furthermore, one treatment in each experiment was selected to represent contrasting soil and management conditions for crops commonly grown in each area. Soils in all three experiments were sandy with low water holding capacities and low organic matter contents. Maize was grown in two experiments and millet in the other. The two crops in West Africa were rainfed whereas the maize crop in Gainesville, Florida was irrigated. Soil parameters, initial conditions, management details, and cultivar coefficients were measured and used in prior simulation studies by the authors of those studies. Weeds were controlled in each experiment and there was no evidence of pest damage. Table 4.1 summarizes the overall characteristics and weather conditions for the three experiments used in this study.

In the first experiment, maize was grown in 1982 in Gainesville, Florida, (29°41′ N 82°21′ W) with irrigation and high nitrogen input to represent a typical high input production system (Table 4.1). The soil is classified as an Arenic Paleudults Fine Sand and has an average depth of about 180 cm. Soil carbon was 0.64% in the top 20 cm of soil. The experiment consisted of six treatments with different irrigation and nitrogen fertilizer inputs. We used the fully irrigated, high nitrogen fertilizer treatment from this 1982 experiment (Bennett et al. 1989) that is distributed with DSSAT (Jones et al. 2003). Rainfall during the season was high (661 mm), temperature was high, but lowest of the three locations, and the season was 126 days long, the longest of the three locations.

Table 4.1 Site characteristics and season average weather conditions for the three experiments

	Gainesville	Wa	Damari
Location	29°41′ N, 82°21′ W	10°3′ N, 2°30′ W	13°12′ N, 2°14′ E
Altitude, (masl)	54	320	196
Soil			
Classification	Arenic Paleudults	Ferric Lixisol	Kanhaplic Haplustult
Texture	Fine sand	Loamy sand	Loamy sand
Relief	Flat	Flat	Flat
SOC (%)	0.64	0.48	0.15
Soil fertility	Low	Very low	Very low
Runoff potential	Low	Low	High
Seasonal Weather			
Rainfall (mm)	661	738	550
Solar radiation (MJ/m²/day)	18.4	18.1	21.4
Max Temp (°C)	29.2	30.2	32.8
Min Temp (°C)	15.8	21.6	23.1
Management			
Planting	Tilled	Tilled	Flat
Fertilizer	400 kg ha^{-1} applied	No N, 39 kg ha^{-1} of P	3,000 kg ha^{-1} manure
Irrigation	Yes	No	No

The second experiment involved a maize trial conducted in 2004 by J. B. Naab (Naab 2005; Naab et al. 2008) in Wa, Ghana, using a treatment that had no nitrogen added but adequate P and other inputs (Table 4.1). The experiment site was located in the Upper West Region of Ghana (10°3′ N, 2°30′ W, altitude 320 m above sea level) and has an unimodal rainfall pattern. The average annual rainfall is 1,100 mm falling mainly between April and September. The mean annual temperature in Wa is 27°C. The soil in Wa is characterized as a Ferric Lixisol in the FAO (2001) classification system, with a loamy sand texture, and having a depth of 60 cm. The organic carbon is very low (0.38% in the top 20 cm soil) and available P of 3.26 mg kg^{-1}. This maize experiment was conducted to evaluate maize response to N and P fertilizer applications using 9 treatments in a factorial experiment design. Details of this study are reported in Naab (2005). The treatment with no N and high P fertilizer input (39 kg ha^{-1} of P) was used in this study. This cropping system represents one with low soil N fertility, no N inputs, with high rainfall. Rainfall during the growing season in Wa was 738 mm in 2004, highest among the three sites, and the fields were nearly flat with low surface runoff potential. Although daily maximum temperature average during the season was only 1°C higher than Gainesville, minimum temperature averaged about 6°C higher. The growing season was 98 days long in Wa, 28 days less than in Gainesville.

The third crop was a flat-planted millet treatment in an experiment at Damari (Niger) in 1999 (Table 4.1) with the application of 3,000 kg ha^{-1} manure (Fatondji et al. 2006). Damari is located at Lat 13°12′ N and Long. 2°14′ E, 45 km from Niamey, the capital city of Niger. The long-term average annual rainfall at Damari is 550 mm, which falls between June and September. The long term monthly average minimum and maximum temperatures vary, respectively, between 16°C in January and 28°C in April and May and between 32°C in January and 42°C in April and May. During the experiment in 1999, weather conditions followed this trend; total rainfall for the season was 499 mm (Table 4.1).

The soil at Damari is classified as a Kanhaplic Haplustult (Soil Survey Staff 1998). Table 4.1 shows measured soil properties (0–60 cm) at the experiment site. The soil is highly acidic (pH-H$_2$O = 3.6–4.5), with 84% sand content, with low effective cation exchange capacity (ECEC) (2.8 cmol kg^{-1}), and very low soil water holding capacity. Because of the soil properties and intense rainfall events in the region the soils are prone to surface crusting (Casenave and Valentin 1989) and high runoff. The soil organic carbon ranged from 0.04% to 0.14% (Fatondji et al. 2006), even lower than the typical levels in Niger (about 0.22%, Bationo et al. 2003) and severely limit yield compared to the genetic potential of the site. Table 4.1 (see Fatondji et al. 2012, this volume) summarize soil parameters for the field in which the Damari experiment was conducted. Despite these extreme conditions, farmers are forced to use these soils for producing crops because of limited land availability. Water harvesting technologies are therefore used to assure better soil water conditions for the crop. Treatments in the experiment were combinations of water harvesting using the zai technology in which seeds are planted in pits (Fatondji et al. 2006) vs. flat planting in combinations with nutrient additions (none, straw residue, and manure). In this study, we selected the flat planted, manure treatment in which either water, nutrients, or both could limit crop production. For other details on the experiment and site, see Fatondji et al. (2006).

DSSAT Cropping System Model

The DSSAT version 4.02 Cropping System Model (CSM) (Jones et al. 2003; Hoogenboom et al. 2004) was used to simulate maize or millet in the experiments described above. This model includes the CERES plant growth models (Ritchie et al. 1998; Ritchie and Alagarswamy 1989) and a dynamic soil water, carbon, and nutrient model (Ritchie 1998; Gijsman et al. 2002; Godwin and Singh 1998; Jones et al. 2003; Dzotsi et al. 2010; Porter et al. 2010) that computes daily changes in the status of soils in response to tillage, irrigation, and nutrient applications and the effect of those soil conditions on crop growth and yield. DSSAT is a software environment that embeds the CSM, and this system has been widely used in research on cropping systems analysis for different purposes.

This model was used by researchers in each of the studies that are included in this analysis. In each study, the soil, weather, and management inputs were put into the model, and cultivar parameters were estimated using the crop development, growth, and yield data from the experiments. For example, Dzotsi et al. (2010) and Naab (2005) used the high N and high P treatment in the Wa (Ghana) experiment to estimate cultivar parameters for the Obatanpa maize variety. Fatondji et al. (2012 this volume) estimated cultivar parameters for the local millet variety used in the Damari experiment. These parameters are given as default values in Tables 4.2, 4.3 and 4.4.

Table 4.2 Distributions of the selected parameters for sensitivity analysis for the maize experiment in Gainesville, FL, USA

Parameter	Parameter Type	Distribution	Statistic Default	STDEV	Min	Max	Description
P1	Cultivar	Uniform	265	–	245.1	284.9	Degree days (base 8°C) from emergence to end of juvenile phase
P2	Cultivar	Uniform	0.3	–	0	0.6	Photoperiod sensitivity coefficient (0–1.0)
P5	Cultivar	Uniform	920	–	851	989	Degree days (base 8°C) from silking to physiological maturity
G2	Cultivar	Uniform	990	–	841.5	999.0	Potential kernel number
G3	Cultivar	Uniform	8.5	–	7.22	9.78	Potential kernel growth rate mg/(kernel days)
PHINT	Cultivar	Uniform	39	–	36.08	41.92	Degree days required for a leaf tip to emerge (phyllochron interval) (°C days)
PAW	Soil water	Normal	0.061	0.0073	–	–	Plant available water, volume fraction
PEW	Soil water	Normal	0.2	0.024	–	–	Potential excess water, volume fraction
SLRO	Soil water	Uniform	65	–	60.1	69.9	Soil runoff curve number
SH2O	Soil water	Uniform	0.086	–	0.05	0.11	Initial condition soil water content, volume fraction
SNH4	Soil nutrients	Uniform	0.5	–	0.3	1.5	Initial condition soil ammonium content, mg kg^{-1}
SNO3	Soil nutrients	Uniform	0.1	–	0.05	0.3	Initial condition soil nitrate content, mg kg^{-1}
SAOC	Soil nutrients	Uniform	0.7	–	0.5	1	Total soil carbon, g/100 g
SASC	Soil nutrients	Uniform	0.85	–	0.8	0.95	Stable organic carbon, fraction
ManureN	Management	Uniform	2.53	–	1.85	3.25	Organic nitrogen, percent dry weight basis
FertilizerN	Management	Normal	400	20	–	–	Inorganic fertilizer nitrogen, kg ha^{-1}

Table 4.3 Distributions of the selected parameters for sensitivity analysis for the maize experiment in Wa, Ghana

Parameter	Type	Distribution	Statistic				Description
			Default	STDEV	Min	Max	
PI	Cultivar	Uniform	280	—	259	301	Degree days (base 8°C) from emergence to end of juvenile phase
P2	Cultivar	Uniform	0	—	0	0	Photoperiod sensitivity coefficient (0–1.0)
P5	Cultivar	Uniform	700	—	647.5	752.5	Degree days (base 8°C) from silking to physiological maturity
G2	Cultivar	Uniform	550	—	467.5	632.5	Potential kernel number
G3	Cultivar	Uniform	7.74	—	6.58	8.9	Potential kernel growth rate mg/(kernel days)
PHINT	Cultivar	Uniform	40	—	37	43	Degree days required for a leaf tip to emerge (phyllochron interval) (°C days)
PAW	Soil water	Normal	0.068	0.0082	—	—	Plant available water, volume fraction
PEW	Soil water	Normal	0.231	0.0277	—	—	Potential excess water, volume fraction
SLRO	Soil water	Uniform	61	—	56.4	65.6	Soil runoff curve number
SH2O	Soil water	Uniform	0.183	—	0.143	0.223	Initial condition soil water content, volume fraction
SNH4	Soil nutrients	Uniform	0.5	—	0.25	0.75	Initial condition soil ammonium content, mg kg^{-1}
SNO3	Soil nutrients	Uniform	1.7	—	0.85	2.55	Initial condition soil nitrate content, mg kg^{-1}
SAOC	Soil nutrients	Uniform	0.48	—	0.4	1.0	Total soil carbon, g/100 g
SASC	Soil nutrients	Uniform	0.85	—	0.5	0.95	Stable organic carbon, fraction
ManureN	Management	Uniform	0	—	—	—	Organic nitrogen, percent dry weight basis
FertilizerN	Management	Normal	0	—	—	—	Inorganic fertilizer nitrogen, kg ha^{-1}

Table 4.4 Distributions of the selected parameters for sensitivity analysis for the millet experiment in Damari, Niger

Parameter	Type	Distribution	Statistic				Description
			Default	STDEV	Min	Max	
P1	Genetic	Uniform	170.0	–	144.5	195.5	Degree days (base 8°C) from emergence to end of juvenile phase
P20	Genetic	Uniform	12.0	–	11.8	12.2	Critical photoperiod or the longest day length (in hours) at which development occurs at a maximum rate
P2R	Genetic	Uniform	150.0	–	127.5	172.5	Extent to which phasic development leading to panicle initiation is delayed for each hour increase in photoperiod above P_2O.
P5	Genetic	Uniform	450.0	–	382.5	517.5	Thermal time (degree days above a base temperature of 10°C from beginning of grain filling (3–4 days after flowering) to physiological maturity)
G1	Genetic	Uniform	1.00	–	0.7	1.3	Scaler for relative leaf size
G4	Genetic	Uniform	0.770	–	0.54	1.00	Scaler for partitioning of assimilates to the panicle (head)
PHINT	Genetic	Uniform	43.00	–	36.55	49.45	Degree days required for a leaf tip to emerge (phyllochron interval) (°C days)
PAW	Soil water	Normal	0.036	0.00432	–	–	Plant available water, volume fraction
PEW	Soil water	Normal	0.301	0.03612	–	–	Potential excess water, volume fraction
SLRO	Soil water	Uniform	98	–	90	100	Soil runoff curve number
SH2O	Soil water	Uniform	0.086	–	0.05	0.11	Initial condition soil water content, volume fraction
SNH4	Soil nutrients	Uniform	0.010	–	0.3	1.5	Initial condition soil ammonium content, mg kg^{-1}
SNO3	Soil nutrients	Uniform	0.007	–	0.05	0.3	Initial condition soil nitrate content, mg kg^{-1}
SAOC	Soil nutrients	Uniform	0.150	–	0.5	1	Total soil carbon, g/100 g
SASC	Soil nutrients	Uniform	0.920	–	0.8	0.95	Stable organic carbon, fraction
ManureN	Management	Uniform	2.530	–	2.03	3.03	Organic nitrogen, percent dry weight basis
FertilizerN	Management	Normal	0	–	–	–	Inorganic fertilizer nitrogen, kg ha^{-1}

Global Sensitivity Analysis

We conducted a global sensitivity analysis on the three systems described above with contrasting climate, soil, and management inputs. We also quantified the uncertainties in yield predictions associated with 16 soil water, soil nutrient, and cultivar parameters.

Parameters in the Sensitivity Analysis

Parameters for sensitivity analysis were selected based on past experiences in adapting crop models to a wide range of soils, climates, and management conditions. Parameters that are usually missing when adapting crop models for a new location are those associated with the cultivars grown there – the cultivar parameters in the DSSAT CSM. Thus, a basic requirement in new situations is to perform experiments and measure crop development, growth and yield to calibrate or estimate the cultivar parameters. Although these parameters should be estimated using data on crops grown under non-limiting resource conditions such as adequate nutrient and water supply with minimal pest damage (Hunt et al. 1993; Boote et al. 2003), this is often not the case. In most cases, crop data are only available for sub-optimal and rainfed trials. The estimation of crop cultivar parameters using such data from crops grown under nutrient or water deficits may not be reliable and would contribute to uncertainty in those parameters. Even though the default parameters for maize and millet listed in Table 4.2 are those reported by researchers who performed the experiments, inherently there are uncertainties associated with these values.

The second set of parameters selected for sensitivity and uncertainty analyses was for the soil water balance, which computes daily amounts of water available in the root zone for crop uptake. Even though researchers may collect soil samples and determine water retention properties in the laboratory, these lab-measured estimates of field water holding characteristics may be inadequate for use in the model because they often fail to capture the field-scale spatial heterogeneity (Ritchie 1998). In addition, model simulation of soil water infiltration is based on a widely-used runoff curve number technique (Williams 1991) that uses the curve number (SLRO) as its defining parameter. This parameter and its determination are highly empirical, and thus its estimates are highly uncertain. Two other parameters that we selected for study are based on the lower limit of water below which plants are not able to extract water (SLLL), the drained upper limit (SDUL), and saturated soil water content (SSAT), and initial soil water content. These parameters are plant available water (PAW, which is (SDUL-SLLL)) and water storage capacity above SDUL (PEW, which is (SSAT-SDUL)).

A final set of five parameters were those associated with soil fertility and its management. Four of the parameters were initial condition estimates for ammonium, nitrate, total soil carbon, and stable soil carbon. Under non-fertilized crops, these factors are very important determinants of nutrient supply during their growing season. A fifth parameter is the input amount of fertilizer N (either inorganic or organic).

Default values and ranges of uncertainty were selected for the 16 parameters. The default values were those previously reported by researchers who used the models to study each of the three experiments; these were available in the DSSAT v4.02 (Hoogenboom et al. 2004) data files. Uniform distributions were used for most parameters. Ranges for the cultivar parameter distributions were based on prior experience in uncertainties obtained when estimating coefficients using field data. Although cultivar parameter uncertainties were not known for the cultivars in these studies, we used the same uncertainty ranges for each location so that differences in sensitivities among the locations could be attributed to location differences instead of differences in parameter uncertainties. Uncertainties in PAW and PEW soil water parameters were described by normal distributions based on a study by He (2008). Tables 4.2, 4.3 and 4.4 show the distributions of the selected parameters for three experiment sites in this study.

Global Sensitivity Analysis

The method for the global sensitivity analysis followed that by Sobol (1993), which is similar to an analysis of variance. Multiple sets of parameters were created using Monte Carlo random sampling from the parameter distributions for running the model to produce output responses that were then analyzed. The variances of response variables were decomposed into the contributions from the various input parameter variations over their ranges of uncertainties (Monod et al. 2006). With 16 parameters or factors, the decomposition of the total variance $\mathrm{var}(\hat{Y})$ in any response Y, such as grain yield, can be summarized by:

$$\mathrm{var}(\hat{Y}) = \sum_{i=1}^{16} D_i + \sum_{i<j} D_{ij} + \ldots + D_{1\ldots16} \tag{4.1}$$

where D_i is the variability associated with the main effect of parameter i, and D_{ij} is the variability associated with the interaction between parameters i and j. Sensitivity indices (S_i) are derived from the decomposition of total variance in Eq.4.1 by dividing the variance attributed to uncertainty in each parameter by $\mathrm{var}(\hat{Y})$:

$$S_i = D_i / \mathrm{var}(\hat{Y}) \tag{4.2}$$

Interactive sensitivity indices can also be computed if needed, based on the D_{ij} terms in Eq. 4.1. In our case, we computed the main effect indices for each parameter along with the total sensitivity, TS_i, to each parameter, i, considering its interactive effects with other parameters, given by:

$$TS_i = \frac{D_i + D_{i2} + \cdots + D_{i\cdots16}}{\mathrm{var}(\hat{Y})} \tag{4.3}$$

The software package SimLab v2.2.1 (Saltelli et al. 2004; SimLab 2005), designed for multiple model runs with probabilistically selected model inputs using Monte Carlo sampling of distributions, was coupled with the DSSAT CSM model to perform the global sensitivity analysis.

The DSSAT-Maize and Millet model runs were executed using each randomly generated sample of input parameters. Distributions of simulated biomass and grain yield were generated, and first order and total sensitivities of these outputs to each uncertain input parameters were computed using the Sobol decomposition of variances. This method requires $N(16+1)$ model runs for the calculation of the first-order sensitivity indices of 16 factors, where N is the number of randomly sampled parameter scenarios. We used $N = 2,048$, which resulted in a total of 34,816 sample sets of input parameters for each site. The parameter conversion and automatic model running were implemented using the R language (R Development Core Team 2009). In essence, these model runs create a mapping from the distribution of parameter uncertainties to the distribution of output uncertainties. We then used the results of these model runs to determine (1) the uncertainty in model predictions for each site, and (2) the input variables with uncertainties that contributed most to yield prediction uncertainty.

Results and Discussion

Simulated grain yield varied with input levels in all the three experiments. For Florida, where the maize crop was fertilized and irrigated, the yield varied between 10,000 and 14,000 kg ha^{-1}, with 12,000 kg ha^{-1} being the most likely model outcome (Fig. 4.1). Maize yield under rainfed conditions and with no N input at Wa (Ghana) varied between 400 and 5,000 kg ha^{-1} but a yield of 1,500 kg ha^{-1} was most likely. In Niger, simulated millet yields varied between 350 and 1,700 kg ha^{-1} with a modal yield of 800 kg ha^{-1}. The fact that the modal yields from the uncertainty analysis were near the observed yields indicates that the model, as calibrated by researchers who conducted those experiments, were about the same as observed. Observed mean grain yields for the treatments used in the experiments from Gainesville, Florida, Wa, Ghana, and Damari, Niger were 11,881, 417, and 705 kg ha^{-1}, respectively. The variabilites in yields shown in Fig. 4.1 were due to the uncertainties in parameters as defined in Tables 4.2, 4.3 and 4.4. Furthermore, uncertainties in yield were higher in absolute terms in Gainesville and Niger, as seen in the spread of these two yield distributions (Fig. 4.1), but the ratios of variances to means were much higher in Ghana and Niger than in Gainesville. One of the main simulation results is that even under a given set of weather conditions at a given location, a range of yields can be realized, primarily due to the variability of inputs. These types of model outputs may provide a more realistic representation of the variable yield outcomes commonly observed on farmers' fields.

Figure 4.2 shows the fractions of total variability in yield that were due to uncertainties in the 16 parameters for the intensive management system at the Gainesville site.

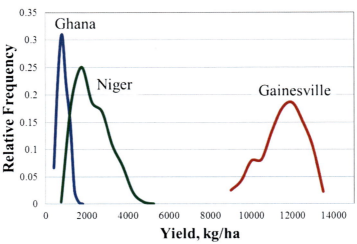

Fig. 4.1 Uncertainty in simulated grain yield for three locations (Gainesville, Florida; Wa, Ghana; and Damari, Niger) that had sandy soils with low nutrient and low water holding capabilities. This graph shows probability distributions that were simulated when taking into account uncertainties in cultivar, soil water, and soil nutrient parameters

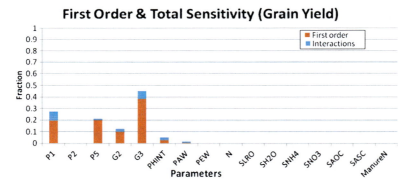

Fig. 4.2 Global sensitivity indices for the Gainesville, Florida site showing the fraction of total grain yield uncertainty that was due to uncertainties in each parameter. The first six are cultivar parameters, which dominated the uncertainty in yield at this site

The lower bars in this stacked-bar figure show first order sensitivity indices Eq. 4.2 for the parameters whereas the very top of the bar shows total sensitivities to each factor Eq. 4.3. Sensitivities to cultivar coefficients were high at this site (accounting for about 90% of the uncertainty). In contrast, sensitivity indices for water and nutrient parameters were very low, accounting for less than 10% of the final yield uncertainty. This is not a surprising result because intensive management provided water and nutrients that were high enough in Gainesville to obscure most effects of

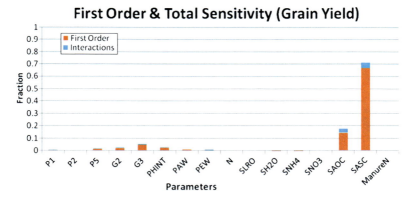

Fig. 4.3 Global sensitivity indices for the Wa, Ghana site showing the fraction of total grain yield uncertainty that was due to uncertainties in each parameter. The two soil carbon parameters dominated the uncertainty in yield at this site

variability in each of the parameters and inputs that would limit yield. In essence, yields in this site were high for all combinations of water and nutrient-related parameters, varying mostly due to cultivar coefficients.

Results were considerably different for the Wa, Ghana location where observed maize yield was 417 kg ha^{-1} (Fig. 4.3). At this site with very low soil nutrient levels, N was mostly supplied by mineralization of soil organic matter during the growing season. Thus, the amount of carbon in the soil and its level of stability were the dominant parameters. As total soil carbon (SAOC) and stable soil carbon (SASC) varied over their levels of uncertainty, mineralization of N for plant growth varied considerably and yield was influenced accordingly. It was somewhat surprising that the effect of stable carbon was higher than total soil carbon. However, if SASC is a large fraction of the total soil C, then very little mineralization would occur even for relatively high SAOC values for these sandy soils. At this site, over 80% of the total yield variability was due to uncertainties in soil carbon characteristics (SAOC and SASC), mostly due to first order effects. The remaining yield uncertainty (less than 20%) was mostly due to cultivar parameters.

In Damari, Niger, water was the most limiting factor for the treatment used in this study, which received 3,000 kg ha^{-1} of manure. Rainfall was lower and runoff was high due to soil crusting. This was clearly shown by the sensitivity factors. The soil water holding capacity (PAW) was the parameter with the highest first order sensitivity whereas the runoff coefficient (SLRO) was second. Together, the water parameters accounted for about 55% of the yield uncertainty. One very interesting result was the interactive effects of soil water and soil nutrient parameters (Fig. 4.4). Although the first order sensitivities of yield to nutrient parameters were very low (less than 1%), the interactive effects of all of the nutrient parameters were high. Considering these interactions, over 85% of the uncertainty of millet grain yield was accounted for. Sensitivities to cultivar parameters were also very low in this site, accounting for less than 13% of the simulated yield variability.

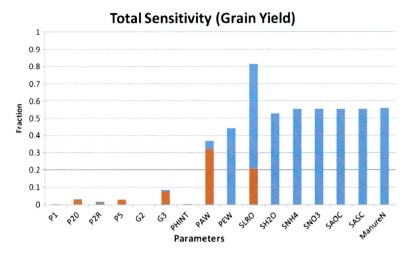

Fig. 4.4 Global sensitivity indices for the Damari, Niger site showing the fraction of total grain yield uncertainty that was due to uncertainties in each parameter. The soil water holding and runoff parameters (*PAW* and *SLRO*) had the highest first order influence on simulated uncertainty in yield at this site. Nutrient parameters also interacted with other parameters when considering total sensitivities

A comparison of sensitivities among sites shows that uncertainties in parameters of factors that limit yield in each site had the highest influence on uncertainties in simulated yields. This information should be considered by those who use cropping system models at locations where yields are low due to one or more limiting factors. Accurate input parameters are needed for those yield-limiting factors. Otherwise, uncertainties in their values can have major effects on uncertainties in simulated yields and other model outputs. This study showed cases where cultivar parameters, soil water parameters, and nutrient parameters can dominate uncertainties in simulated yields, depending on the production situation being studied. Another implication of these findings is that one should not estimate cultivar parameters using observed crop growth variables from studies in which soil limitations restrict crop growth. Varying these cultivar parameters may have little or no effect on simulated crop growth and yield results due to water, nutrient, or other factors that may severely limit growth and yield.

This study showed that sensitivities of simulated yields to soil parameters can dominate those for cultivar parameters in degraded soils and low input cropping systems. In effect, much more attention needs to be paid to the determination of input soil properties than has hitherto been the case. Undoubtedly, the determination of soil properties often entails high costs and is often time consuming. However, recent advances in measurement techniques using simple field soil testing kits should provide a feasible means for data collection in variable landscapes for use in modeling work.

Conclusions

Contributions of factors to simulated overall yield uncertainties in three contrasting production situations were expressed by sensitivity indices in this chapter. Sensitivities due to cultivar parameters were high under intensive management in Florida and low in harsh environments (Ghana and Niger). We also concluded that sensitivities to soil water parameters were low under intensive management (Florida) and nutrient-limited (Ghana) environments, but high in degraded soil of Niger when nutrients were supplied via manure. Sensitivities to C and N parameters were low in Florida and Niger when nutrients were supplied but high in Ghana when no nutrients were applied. Our results showed that some parameters may not have high first order sensitivities yet have major influences on model outputs via interactions with other factors. Sensitivities to soil parameters dominate those for cultivar parameters in degraded soils and low input cropping systems. This study also showed that some parameters may not have high first order sensitivities yet have major influences on model outputs via multi co-linearity and interactions with other factors.

In low input farming systems, other uncertainties that were not considered in this study are likely to be dominant in some situations. In particular, biotic stresses caused by weed competition, plant diseases and insect damage may greatly influence yield, and there are inherent uncertainties in the type, magnitude, and timing of biotic stresses due to the difficulties in measuring and modeling these yield-reducing factors. Uncertainty and sensitivity analysis methods used in this paper to study uncertainties in cultivar and soil inputs can also be used for those factors if the distributions of these factors can be estimated. This research further highlights the need for more attention to uncertainties in model predictions under a range of production situations. Global sensitivity analysis is needed to help ensure that field-scale parameter estimates are anchored in an understanding of model behavior for specific cropping systems.

References

Bationo A, Mokwunye U, Vlek PLG, Koala S, Shapiro BI (2003) Soil fertility management for sustainable land use in the West African Sudano-Sahelian zone. In: Gichuru MP et al (eds) Soil fertility management in Africa: a regional perspective. Academy Science Publisher & Tropical Soil Biology and Fertility, Nairobi, pp 253–292

Bennett JM, Mutti LSM, Rao PSC, Jones JW (1989) Interactive effects of nitrogen and water stresses on biomass accumulation, nitrogen uptake, and seed yield of maize. Field Crop Res 19:297–311

Boote KJ, Jones JW, Batchelor WD, Nafziger ED, Myers O (2003) Genetic coefficients in the CROPGRO-soybean model: links to field performance and genomics. Agron J 95:32–51

Casenave A, et Valentin C (1989) Les états de surface de la zone sahelienne; Influence sur l'infiltration. Les processus et les facteurs de réorganisarion superficielle. (ed) ORSTOM – Institut Français de Recherche Scientifique pour le Développement en Coopération. Collection Didactiques. Paris, pp 65–190

Dzotsi KA, Jones JW, Adiku SGK, Naab JB, Singh U, Porter CH, Gijsman AJ (2010) Modeling soil and plant phosphorus within DSSAT. Ecol Model 221:2839–2849

FAO (2001) Lecture notes on the major soils of the world. World Soil Res Rep. 289 pp

Fatondji D, Bationo A, Tabo R, Jones JW, Adamou A, Hassane O (2012) (this volume) Water use and yield of millet under the zai system: understanding the processes using simulation

Fatondji D, Martius C, Bielders C, Vlek P, Bationo A, Gérard B (2006) Effect of planting technique and amendment type on pearl millet yield, nutrient uptake, and water use on degraded land in Niger. Nutr Cycl Agroecosyst 76:203–217

Gijsman AJ, Hoogenboom G, Parton WJ, Kerridge PC (2002) Modifying DSSAT crop models for low-input agricultural systems using a soil organic matter-residue module from CENTURY. Agron J 94:462–474

Godwin DC, Singh U (1998) Nitrogen balance and crop response to nitrogen in upland and lowland cropping systems. In: Tsuji GY, Hoogenboom G, Thornton PK (eds) Understanding options for agricultural production. Springer, Dordrecht, pp 55–78

Hammer GL, Muchow RC (1991) Climatic risk in crop production: models and management for the semi-arid tropics and subtropics. CAB International, Wallingford, pp 205–232

He J (2008) Best management practice development with the CERES-maize model for sweet corn production in North Florida. PhD dissertation, University of Florida, Gainesville, 329 pp

Hoogenboom G, Jones JW, Wilkens PW, Porter CH, Batchelor WD, Hunt LA, Boote KJ, Singh U, Uryasev O, Bowen WT, Gijsman AJ, du Toit AS, White JW, Tsuji GY (2004) Decision support system for agrotechnology transfer version 4.0, [CDROM]. University of Hawaii, Honolulu

Hunt LA, Pararajasingham S, Jones JW, Hoogenboom G, Imamura DT, Ogoshi RM (1993) GENCALC: software to facilitate the use of crop models for analyzing field experiments. Agron J 85:1090–1094

Jones JW, Hoogenboom G, Porter CH, Boote KJ, Batchelor WD, Hunt LA, Wilkens PW, Singh U, Gijsman AJ, Ritchie JT (2003) The DSSAT cropping system model. Eur J Agron 18(3–4): 235–265

Matthews RB, Stephens W (2002) Crop-soil simulation models: applications in developing countries. CABI Publishing, Wallingford, 280 pp

Monod H, Naud C, Makowski D (2006) Uncertainty and sensitivity analysis for crop models. In: Wallach D, Makowski D, Jones JW (eds) Working with dynamic crop models: evaluation, analysis, parameterization, and applications. Elsevier, Amsterdam, pp 84–87

Naab JB (2005) Measuring and assessing soil carbon sequestration by agricultural systems in developing countries, 2004 annual report. Savanna Agricultural Research Institute, Wa

Naab JB, Koo J, Traore PCS, Adiku SGK, Jones JW, Boote KJ (2008) Carbon Enhancing Management Systems(CEMS): estimation of soil carbon sequestration potential in smallholder farming systems in Northern Ghana. Technical bulletin 2008–3, ABE Department, University of Florida, Gainesville, 11 pp

Porter CH, Jones JW, Adiku S, Gijsman AJ, Gargiulo O, Naab JB (2010) Modeling organic carbon and carbon-mediated soil processes in DSSAT v4.5. Oper Res Int J 10(3):247–278

R Development Core Team (2009) R: a language and environment for statistical computing. R Foundation for Statistical Computing, Vienna. ISBN 3-900051-07-0

Ritchie JT (1998) Soil water balance and plant water stress. In: Tsuji GY, Hoogenboom G, Thornton PK (eds) Understanding options for agricultural production. Springer, Dordrecht, pp 41–54

Ritchie JT, Alagarswamy G (1989) Genetic coefficients for CERES models. In: Virmani SM, Tandon HLS, Alagarswamy G (eds) Modeling the growth and development of sorghum and pearl millet. ICRISAT research bulletin no. 12. ICRISAT, Patancheru, pp 27–34

Ritchie JT, Singh U, Godwin DC, Bowen WT (1998) Cereal growth, development and yield. In: Tsuji GY, Hoogenboom G, Thornton PK (eds) Understanding options for agricultural production. Springer, Dordrecht, pp 79–98

Saltelli A, Tarantola S, Campolongo F, Ratto M (2004) Sensitivity analysis in practice: a guide to assessing scientific models. Wiley, Chichester

SimLab (2005) SimLab Ver. 2.2. Reference manual

Sobol IM (1993) Sensitivity estimates for non-linear mathematical models. Math Model Comp Exp 1(4):407–414

Soil Survey Staff (1998) Keys to soil taxonomy, 8th edn. USDA/NRCS, Washington, DC

Thornton PK, Wilkens PW (1998) Understanding options for agricultural production: systems approaches for sustainable agricultural development. Kluwer, Dordrecht, pp 329–345

Williams JR (1991) Runoff and water erosion. In: Hanks RJ, Ritchie JT (eds) Modeling plant and soil systems. Agronomy monograph #31, American Society of Agronomy, Madison

.

Chapter 5
The Response of Maize to N Fertilization in a Sub-humid Region of Ghana: Understanding the Processes Using a Crop Simulation Model

D.S. MacCarthy, P.L.G. Vlek, and B.Y. Fosu-Mensah

Abstract Crop simulation models afford the opportunity to study and understand underlying processes that impact on crop yield, hence, helps in designing appropriate strategies to improve crop production. The response of maize to N fertilization in a sub-humid environment was evaluated using DSSAT (crop simulation model). Two field experiments were conducted in the major and minor seasons in 2007. One was conducted under limited water and nutrient stress conditions and data collected used to calibrate model. The second independent experiments were conducted with different levels (0, 40, 80 and 120 kg N ha^{-1}) of N fertilizer. Grain and biomass yields were predicted with an index of agreement of between 0.64 and 0.95 in both major and minor seasons. Biomass N content and crop phenology were also adequately simulated. Model simulations were better with higher rates of N fertilization and lesser water stress conditions. Water stress during the reproductive stage significantly affected grain yield. Sensitivity of selected soil parameters to grain yield indicated more sensitivity when no N fertilizer was applied. The DSSAT model has been satisfactorily calibrated and evaluated for the study area and hence, can be used to aid decision making in respect of farm management options.

D.S. MacCarthy (✉)
College of Agriculture and Consumer Sciences, Institute of Agricultural Research, Kpong Research Centre, University of Ghana, P.O. Box LG 68, Legon, Accra, Ghana
e-mail: kpongor@yahoo.com

P.L.G. Vlek • B.Y. Fosu-Mensah
Center for Development Research (ZEF), University of Bonn,
Walter-Flex-Str. 3, 53113 Bonn, Germany

J. Kihara et al. (eds.), *Improving Soil Fertility Recommendations in Africa using the Decision Support System for Agrotechnology Transfer (DSSAT)*,
DOI 10.1007/978-94-007-2960-5_5, © Springer Science+Business Media Dordrecht 2012

Introduction

The most prominent constraint to food production in most parts of the world is low fertility (Bationo and Mokwunye 1991). Increasing human population pressure has decreased the availability of arable land and it is no longer feasible to use extended fallow periods to restore soil fertility and organic carbon. This fallow period has been reduced to much shorter durations that can no longer regenerate soil productivity (Nandwa 2001). High population densities have necessitated the cultivation of marginal lands that are prone to erosion, hence enhancing environmental degradation through soil erosion and nutrient mining. As a result, the increase in yield has been due largely to land expansion rather than to crop improvement potential. Although food production in some sub regions has been reported to have increased over the years, rapid growth in population has resulted in a decline in the per capita food production. Due to crop intensification the soils that are the resource base for food production are also deteriorating (Bationo 2006). This threatens future food security unless measures are taken to stop this phenomenon.

Nitrogen is the most limiting soil nutrient in cereal production in Ghana. Maize (*Zea mays* L) is an important food staple in and is used widely across the country for various types of foods. Its consumption outweighs that of rice. Unfortunately its cultivation has been affected with many problems among which low soil fertility is a major issue. To further exacerbate this problem, maize is cultivated mainly by small holder farmers who invest very little in the application of inorganic fertilizer.

Identifying major yield limiting factors and appropriate crop management practices would require many years of experimentation in order to be able to make meaningful deductions. This can be expensive and very time consuming. Crop simulation models, however, provide an excellent alternative approach (MacCarthy et al. 2009; Akponikpe et al. 2008). Crop simulation models afford us the opportunity to study the underlying processes that affect yield and crop production. Crop simulation models have been used to access yield gaps in peanut production in the Guinea Savannah of Ghana (Naab et al. 2004). Moeller (2004) used a crop simulation model (CSM) as a tool to analyse the sustainability of wheat-chickpea rotation. Nutrient use efficiency and water productivity were also analysed for a semi-arid region in Ghana (MacCarthy et al. 2010) using CSM. CSMs require information on soil, crop management, crop cultivar specific coefficients and climatic information (daily maximum and minimum temperature, solar radiation and rainfall) to simulate crop and soil processes and to predict yield. They offer the opportunity to study 'what if' type of situations in which various options are compared. The objective of this was to calibrate a medium duration hybrid maize (mamaba) cultivar for the study region and study the response of the Cropping System Model (CSM)-CERES-Maize model to mineral N fertilizer application. Additionally, the processes that underlay the yield response were also studied as to inform its further use in supporting decisions regarding agricultural practices.

Materials and Methods

This study was carried out in a sub-humid agro-ecological zone at Ejura (Sekyedumase district of Ashanti region), in Ghana. Ejura is located within latitude 1°21″ west and longitude 7°22″ north and its climate is characterized by a bi-modal rainfall pattern with the first rainfall (major) season from mid April to August and the second (minor) season from September to November, usually with a brief break (less than a month) between the two seasons. Weather data (daily rainfall amounts, solar radiation, maximum and minimum temperature) used in this study were collected from a weather station located about 2 km from experimental fields. Average annual total rainfall amount is about 1,439 mm and mean annual maximum and minimum temperatures of 33.5°C and 22°C. A summary description of the general nature of the soil in the study area is presented in Table 5.1.

Experiment for Model Calibration

Two experiments were conducted under limited water and nutrient stress conditions for model calibration in the major and minor seasons in 2007. A medium duration maize variety was planted on the 9th of May and on the 1st of September. 120 kg N ha^{-1} in the form of ammonium sulphate was applied in two splits on the 10th and 40th days after planting. Phosphorous (P_2O_5) fertilizer was applied at 60 kg ha^{-1} in the form of triple super phosphate. Phenological data on emergence, anthesis and physiological maturity were collected. Dates were noted when 50% of plant population attained a particular stage. Plant biomass accumulation was monitored regularly over the growing seasons. Total biomass and grain yield were determined from an area of 3 m^2, oven dried to a constant weight and expressed in dry weight on a kg ha^{-1} basis. Additionally, nitrogen content of total dry matter was determined on biomass collected every 3 weeks. Pre-sowing soil sampling was also done from different horizons from a soil profile and analysed for organic carbon, pH, particle size

Table 5.1 General soil chemical and physical characteristics of the study site, Ejura, Ghana

Soil parameter	0–15 (cm)	15–30 (cm)
Soil organic carbon (%)	0.62	0.54
Cation exchange capacity Cmol c kg soil^{-1}	6.52	5.65
pH	5.40	5.80
Available P mg kg^{-1}	6.43	5.49
Total N	0.03	0.02
Sand (%)	69	61
Silt (%)	13	15
Clay (%)	18	24

Table 5.2 Soil physical and chemical characteristics used in the simulation

Depth (cm)	DUL	LL15	BD	SAT	pH	OC
0–15	0.250	0.081	1.39	0.440	4.8	0.6
15–30	0.287	0.157	1.44	0.428	5.9	0.52
30–60	0.350	0.208	1.51	0.400	5.8	0.45
60–80	0.350	0.250	1.49	0.400	6.0	0.45
80–100	0.350	0.250	1.49	0.400	6.0	0.37

distribution and bulk density. The methods used for these analyses are described in detail in Hoogenboom et al. (1999). Soil moisture characteristics such as field capacity (DUL), permanent wilting point (LL) and saturated moisture content of soils were determined using a pedo-transfer function.

Experiment for Model Evaluation

Experimental design for this study was a randomized complete block design with three replicates. A medium duration maize variety, Mamaba was used. Four levels (0, 40, 80, 120 kg N ha^{-1}) of N fertilizer were applied as treatments. Additionally, P fertilizer was also applied to all plots at a rate of 60 kg P_2O_5 ha^{-1}. Experimental trials were carried out in 2007. Planting in the major season was on the 9th of May and on the 1st September in the minor season. The N fertilizer was split applied on 10 and 40 days after planting in both seasons, while the P fertilizer was applied only on the 10th day after sowing. The plot sizes of 6 m × 6 m were used with a plant spacing of 75 cm × 40 cm.

Data on plant phenological stages (date of seed emergence, date of end of juvenile stage, date of anthesis and date of physiological maturity) were collected. Dates were noted when 50% of plant population attained a particular stage. Plant biomass accumulation was monitored regularly over the growing seasons (every 3 weeks). Total biomass and grain yield were determined from an area 3 m^2, oven dried to a constant weight and expressed in dry weight on kg ha^{-1} basis. Pre-sowing soil samples (disturbed and undisturbed) were collected from soil profiles from different horizons (Table 5.2) and analysed for organic carbon, pH (water), mineral N, CEC, particle size distribution and bulk density. The wilting point, field capacity and saturation moisture content were derived using a pedo-transfer function.

Model Description

The CSM-CERES-Maize module of the Decision Support System for Agro-technology Transfer (DSSAT version 4.0) was used in this study. It describes the dynamics of plant growth and development, soil water, soil carbon, soil nitrogen as a function of

Table 5.3 Genetic coefficients of the maize (mamaba) cultivar used for model simulations

Definition	Abbreviation	Value
Thermal time from seedling emergence to the end of the juvenile	P1	220
Extent to which crop development (expressed as days) is delayed for each hour increase in photoperiod above the optimal photoperiod	P2	0.00
Thermal time from silking to physiological maturity (expressed in degree days above a base temperature of 8°C)	P5	630
Maximum possible number of kernels per plant	G2	850
Kernel filling rate during the linear grain filling stage and under optimum conditions (mg/day)	G3	7
Phylochron interval; the interval in thermal time (degree days) between successive leaf tip appearances	PHINT	42

weather parameters, crop genetic information (genetic coefficients), crop and soil management and cropping history. Phenological development is a function of growing degree days or thermal time and photoperiod. The CERES-Godwin based soil organic matter module was used in this study. The carbon and nitrogen modules simulate mineralization and immobilization of mineral N, nitrification of ammonium, denitrification of nitrate and hydrolysis of urea. The soil water simulation is based on the cascading soil water balance method. Soil water characteristics are specified in terms of drained upper limit (DUL), lower limit (LL) plant extractable soil water content and saturated water content (SAT). Surface evaporation is simulated using the Ritchie (1998) approach. Runoff from rainfall and irrigation was estimated based on the empirical USDA curve number approach. The model uses the approach of Priestley and Taylor (1972) for calculating potential evapo-transpiration.

Model Calibration and Evaluation

The mean estimated parameters of the two seasons (under optimal growth conditions) were used to calibrate the CSM-CERES – Maize module (Jones et al. 1998). Thermal time is computed using an algorithm by Jones and Kiniry (1986) which assumes development rate increases as a linear function of temperature between the base temperatures (8°C) and an optimal temperature of 34°C. Based on the phenological data collected, genetic coefficients (P1, P5 and PHINT) were calculated from daily temperature data collected for the study area as mentioned earlier. These coefficients were then fine-tuned to attain appreciable agreement between simulated and observed values for the anthesis and physiological maturity stages. Table 5.3 shows the five cultivar coefficients and their values that were used in the study. The genetic coefficients for G2 and G3 were determined by iteration of model simulations based on data collected under limited growth stress condition. Iterations were repeated until there was an appreciable agreement between simulated and observed values for the yield data.

Statistical Analysis

Analysis of variance was carried out to determine grain and biomass yield response to applied nitrogen. T-test pair-wise means comparison was employed to determine significant difference between simulated and observed values. The performance of the model in simulating grain and biomass yield were assessed using the root mean square error (RMSE) and index of agreement (d) (Willmott et al. 1985; Loague and Green 1991). RMSE is defined as;

$$\text{RMSE} = \left[n^{-1} \sum \left(Yield_{Simulated} - Yield_{Observed} \right)^2 \right]^{0.5}$$

$$d = 1 - \left[\frac{\sum_{i=1}^{n} (S_i - O_i)(S_i - O_i)}{\sum_{i=1}^{n} (|S_i - M_i| + |O_i - M_i|)} \right]$$

where S_i and O_i are the simulated and the observed values, n the number of observations, and M the mean of the observed value. It is expected for a good simulation to have values of RMSE and d as close as possible to 0 and 1 respectively. High values of d close to 1 indicate good model performance and better relation of observed verses simulated.

Sensitivity Analysis

The sensitivity of grain yield to selected soil parameters were carried out in both major and minor seasons by considering changes of +75%, +50%, +25%, 0%, −25%, −50%, −75%. This was done for no fertilizer treatment and treatment with 80 kg N ha^{-1}. The selected parameters were organic carbon, pH, total N and bulk density. Changes in grain yield were also expressed in percentages.

Results and Discussions

Rainfall, Temperature and Solar Radiation, and Soil

The total rainfall amount in 2007 for the in-crop duration were 272 mm and 397 mm for the major and minor seasons, respectively (Fig. 5.1), compared with an annual amount of 1,300 mm. Maximum daily temperature in the major season ranged from 29°C to 36°C with an average value of 32°C, while minimum daily temperature ranged between 21°C and 34°C with an average value of 23°C. Average solar radiation

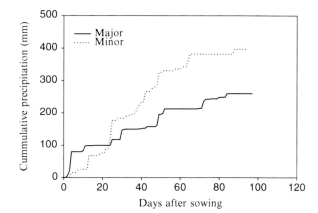

Fig. 5.1 Cumulative rainfall amounts for the major and minor season in 2007 at Ejura, in Ghana

was 18 MJ m^{-2} with a minimum value of 7 MJ m^{-2} and maximum value of 25 MJ m^{-2} in the major season. In the minor season, maximum daily temperature ranged from 30°C to 35°C with an average value of 33°C, while minimum daily temperature ranged from 19°C to 24°C with an average of 22°C. Daily solar radiation ranged from 7 to 24 MJ m^{-2} with a daily average of 19 MJ m^{-2} in the minor season. The two seasons were similar in terms of the minimum and maximum temperatures as well as solar radiation. However, they differed in terms of the total rainfall amounts received, hence relevant to test the models capability to response to nitrogen fertilization under varied rainfall amounts.

There was a significant difference ($p < 0.05$) between depths for all soil parameters (Table 5.1). The soils had a percentage of sand above 60%. The pH$_{(water)}$ was acidic for all plots and depths. The soil was low in organic carbon, total N, and available P, decreasing with depth. The cation exchange capacity was also low.

Model Calibration

The total number of days from sowing to anthesis was 53, which is equivalent to 1063 thermal degree days with a root mean square error of 1 day in the calibration experiments. Physiological maturity was attained on 95 days after sowing with a RMSE of 3 days. 1,889 thermal degree days was required to attain physiological maturity. Total biomass accumulation measured was 12.9 t ha^{-1} with a RMSE of 0.65 t ha^{-1}. Grain yield measured was 4.3 t ha^{-1} with RMSE of 0.29 t ha^{-1}. Number of days taken for anthesis to occur, to attain physiological maturity, total biomass and grain yield were all reasonably predicted. Additionally, nitrogen content of biomass over the calibration experiment in the minor season (under limited nutrient and water stress) was well simulated over the entire growth period with a RMSE of 15 kg N ha^{-1} and an r^2 value of 0.89.

Fig. 5.2 Cumulative thermal degree time of maize grown for the major and minor season at Ejura, Ghana. *Arrows* indicate emergence (e), anthesis (f1, f2) for the minor and major season respectively, and physiological maturity (m1, m2)

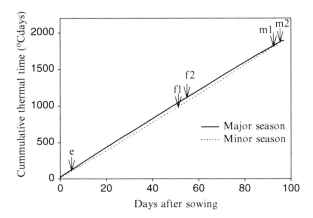

Evaluation of Crop Growth, Grain and Total Biomass

The evaluation of the CSM-CERES-Maize model for simulating the duration from sowing to anthesis with data from both major and minor season trials revealed similar average values between observed and predicted values of 58 days. The coefficient of determination (r^2) between the simulated and observed duration from planting to anthesis was 0.89, with the slope of the regression equation not statistically different from one and the intercept not different from zero ($P = 0.05$). The CSM-CERES-Maize model adequately simulated the duration from sowing to physiological maturity at 95 days with a RMSE of 3 days. A T-test analysis revealed that simulated and observed physiological maturity were not statistically different ($p = 0.05$), confirming the ability of the CSM-CERES-Maize model for simulating the duration from sowing to physiological maturity. The differences in model predictions of days to anthesis and maturity between the calibration and validation without adjustment of model parameters after calibration are an indication that the model is sensitive to the changing environmental conditions. However, the effect of N stress is not well reflected.

The delay in anthesis and maturity in the major season (Fig. 5.2) is probably due to the water stress effect on crop growth and photosynthesis as indicated by the simulation in Fig. 5.3. Gungula et al. (2003) reported similar results on their study on predictions of maize phenology under nitrogen-stressed conditions in Nigeria. Moisture stress will result indirectly in nitrogen stress as water is needed for nutrient uptake. Water stress was more prominent at higher levels of fertilization compared to when no fertilizer was applied (Fig. 5.3) since fertilization leads to higher biomass production which in turn will require more resources to sustain the accumulated biomass. The uptake of nitrogen was satisfactorily represented by model though predictions at 80 kg N ha^{-1} treatment in the minor season were over predicted by the model. On the contrary, N uptake was underestimated under no fertilizer application treatment (Fig. 5.4).

Fig. 5.3 Simulated effect of soil water stress factor on growth for selected treatments of the experiment conducted in the major and minor season

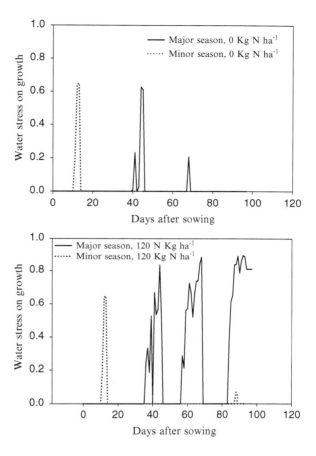

Fig. 5.4 Comparison of simulated and observed nitrogen uptake by the maize crop biomass in the minor season of 2007 at Ejura, Ghana

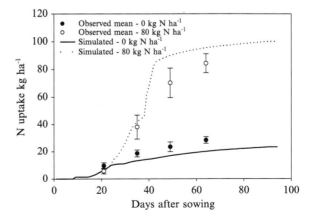

Grain and biomass yield for the major and minor seasons are given in Table 5.4. Both grain and biomass yield were higher in the minor season than the major season. Grain yield ranged from 1.0 to 3.9 $t\,ha^{-1}$ in the major season, and from 1.2 to 4.0 $t\,ha^{-1}$ in the minor season. Biomass yield also ranged between 2.5 and 6.3 $t\,ha^{-1}$ in the major season and from 2.8 to 8.9 $t\,ha^{-1}$ in the minor season. The trend of total above ground biomass production was well simulated by the model (Fig. 5.5) Table 5.5.

The model predicted grain and biomass yield to in response to nitrogen fertilization satisfactorily in both major and minor seasons. Predictions were however better in the minor season reflecting some difficulty in simulating water stress properly, with a RMSE ranging from 0.22 to 0.62 $t\,ha^{-1}$ and 0.19 to 0.51 $t\,ha^{-1}$ in the minor season. Model predictions were better at higher N levels than when no fertilizer was applied. Both grain and biomass yield were under predicted at higher magnitudes when no fertilizer was applied compared to fertilizer application. The effect of water and N stress on grain and biomass were generally predicted satisfactorily. Yield in general was higher in the minor season which had a higher growing season total rainfall than the major season. The converse is usually the case, as rainfall amounts are usually higher in the major season. This shows the importance of sowing dates in determining crop yield as this could affect the cumulative rainfall amounts during the growing season. Additionally, both N and water stress had a negative impact on grain yield and biomass particularly in the major season (Figs. 5.3, 5.6 and 5.7). The simulated water stress on growth occurred during the reproductive phase of maize growth in the major season. Since grain or kernel size depends on the biomass at anthesis (Akponikpe et al. 2010) water and nitrogen stress during this particular development phase explain the reduction in grain yield compared to the minor season.

Sensitivity Analysis

Grain yield was not sensitive to any of the soil parameters for the high N fertilization (80 kg N ha^{-1}). The amount of N fertilizer applied was sufficient to mask the effect of varying the selected parameters on grain yield. Similar results were reported by Akponikpe et al. (2010) in their study regarding nitrogen management for pearl millet in the Sahel using APSIM model for long-term simulation to support agricultural decision making. With the control where no fertilizer was applied, grain yield was most sensitive to bulk density and pH. Similar trends were exhibited in both major and minor seasons though the magnitudes varied (Fig. 5.8). The sensitivity of the model in simulating grain yield response to changes in organic carbon was very marginal probably due to the Godwin's organic matter method used in simulating soil organic matter and nitrogen. The use of this method of carbon and nitrogen simulation present some challenges in smallholder farming settings where crop production relies to a large extent on inherent soil fertility and use of organic materials. Using the Century model method of carbon and nitrogen simulation could improve the sensitivity of grain yield to changes in organic carbon content of the soil.

Table 5.4 Comparisons of predicted and observed grain and biomass yield in response to varying N rates in sub-humid region, Ejura, Ghana. Observed values are means with standard deviations in parenthesis

N applied (kg ha^{-1})	Grain yield (t ha^{-1})				Biomass yield (t ha^{-1})			
	Major season		Minor season		Major season		Minor season	
	Simulated	Observed	Simulated	Observed	Simulated	Observed	Simulated	Observed
0	0.84	1.0(0.17)	0.88	1.2(0.24)	2.5	2.6(0.2)	2.7	2.8(0.1)
40	2.3	2.4(0.18)	2.7	2.3(0.26)	5.7	4.7(0.1)	7.0	7.1(0.5)
80	2.9	3.3(0.29)	4.4	3.5(0.30)	6.7	5.8(0.3)	8.4	7.7(0.6)
120	3.6	3.9(0.24)	4.6	4.0(0.34)	6.9	6.3(0.5)	8.7	8.9(0.8)

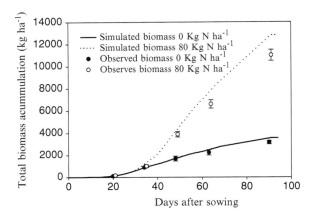

Fig. 5.5 Comparison of simulated and observed biomass accumulation of maize in the minor season of 2007 at Ejura, Ghana

Table 5.5 Root mean square error (RMSE), linear correlation coefficient (R^2) and index of agreement (d) of simulated vs. measured maize grain and biomass grain yield in response to N fertilization in Ejura in the major and minor seasons of 2007. Number of observations (n)

N		Grain yield		Biomass yield	
		Major	Minor	Major	Minor
12	RMSE	0.45	0.43	0.83	0.69
12	d	0.64	0.87	0.93	0.95
12	R^2	0.87	0.87	0.82	0.83

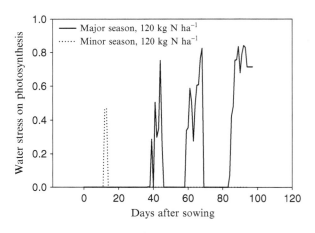

Fig. 5.6 Simulated effect of soil water stress factor on photosynthesis for selected treatments of the experiments conducted during the major and minor seasons

Fig. 5.7 Simulated effect of soil nitrogen stress factor on growth for selected treatments in experiment in the major and minor seasons

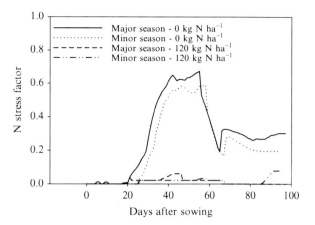

Fig. 5.8 DSSAT sensitivity analysis of maize growth yield to variation in selected soil parameters on treatments with 0 N and 80 kg N ha⁻¹ in the major and minor seasons of 2007 at Ejura, Ghana

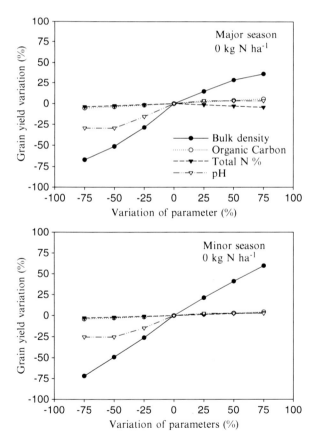

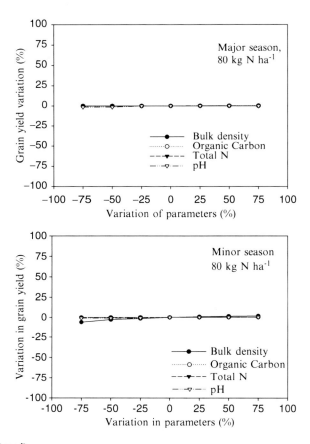

Fig. 5.8 (continued)

Conclusions

The CSM-CERE-Maize model of the DSSAT version 4.0 simulated maize grains and biomass yield adequately in response to inorganic nitrogen fertilization with index of agreement of simulated and observed values of 0.64 (major season grain yield) to 0.95 (minor season biomass yield). RMSE of simulations ranged from 0.22 to 0.62 t ha⁻¹ and 0.19 to 0.51 t ha⁻¹ in the minor season, for instance. Simulations were generally better under nitrogen fertilization than under no fertilization treatment. Model simulations on yields were better under less nutrient and water stress. Results from this study indicate water and nitrogen stress should be avoided during the reproductive stage to maximize grain yield. Thus, in case of limited inorganic fertilizer and water accessibility for irrigation, the reproductive stage needs to be well targeted to maximize grain yield. These results are valid to the extent that phosphorous fertilizers are applied, as this version of the model is not responsive to phosphorous. The model has been adequately evaluated and can therefore be used to aid decision making regarding inorganic nitrogen fertilization as well as water management practices in the study region.

References

Akponikpe P, Michels K, Bielders C (2008) Integrated nutrient management of pearl millet in the Sahel combining cattle manure, crop residue and mineral fertilizer. Exp Agric 44:453–472

Akponikpe P, Gérard B, Michels K, Bielders C (2010) Use of the APSIM model in long term simulation to support decision making regarding nitrogen management for pearl millet in the Sahel. Eur J Agron 32:144–154

Bationo A (2006) Organic amendments for sustainable agricultural production in Sudano-Sahelian West Africa. In: Bationo A (ed) Integrated soil fertility management options for agricultural intensification in Sudano-Sahelian zone of West Africa. Academy Science, Nairobi

Bationo A, Mokwunye AU (1991) Role of manure and crop residues in alleviating soil fertility constraints to crop production: with special reference to the Sahelian and Sudanian zones of West Africa. Fert Res 29:117–125

Gungula DT, Kling JG, Togun AO (2003) CERES-maize predictions of maize phenology under nitrogen-stressed conditions in Nigeria. Agron J 95:892–899

Hoogenboom G, Wilkens PW and Tsuji GY (eds) (1999) DSSAT version 3 (4). International consortium for agricultural systems application. University of Hawaii, Honolulu

Jones CA, Kiniry JR (1986) CERES-maize: a simulation model of maize growth and development. Texas A&M University Press, College Station

Jones JW, Tsuji GY, Hoogenboom G, Hunt LA, Thornton P, Wilkens PW, Imamura D, Bowen WT, Singh U (1998) Decision support systems for agrotechnological transfer (DSSAT) v3. In: Tsuji GY et al (eds) Understanding options for agricultural production, vol 7. Kluwer Academic, Dordrecht, pp 157–177

Loague K, Green RE (1991) Statistical and graphical methods for evaluating solute transport models: overview and application. J Contam Hydrol 7:51–73

MacCarthy DS, Sommer R, Vlek PLG (2009) Modeling the impacts of contrasting nutrient and residue management practices on grain yield of sorghum (Sorghum bicolor (L.) Moench) in a semi-arid region of Ghana using APSIM. Field Crop Res 113(2):105–115

MacCarthy DS, Vlek PLG, Bationo A, Tabo R, Fosu M (2010) Modeling nutrient and water productivity of sorghum in smallholder farming systems in a semi-arid region of Ghana. Field Crop Res 118(3):251–258

Moeller C (2004) Sustainable management of a wheat-chickpea rotation in a Mediterranean environment: scenario analyses using a cropping system simulator. Agroecology 6. PhD Thesis

Naab JB, Singh P, Boote KJ, Jones JW, Marfo KO (2004) Using the cropgro-peanut model to quantify yield gaps of peanut in the Guinean Savanna zone of Ghana. Agron J 96:1231–1242

Nandwa SM (2001) Soil organic carbon (SOC) for sustainable productivity of cropping and agro-ecosystems in Eastern and Southern Africa. Nutr Cycl Agroecosyst 61:143–158

Priestley CH, Taylor RJ (1972) On the assessment of surface heat flux and evapotranspiration using large scale parameters. Mon Weather Rev 100:81–91

Ritchie JT (1998) Soil water balance and plant water stress. In: Tsuji GY, Hoogenboom G, Thornton PK (eds) Understanding options for agricultural production. Kluwer Academic, Dordrecht, pp 41–54

Willmott CJ, Ackleson GS, Davis RE, Feddema JJ, Klink KM, Legates DR, O'donnell J, Rowe CM (1985) Statistics for the evaluation and comparison of models. J Geophys Res 90:8995–9005

Chapter 6
Water Use and Yield of Millet Under the Zai System: Understanding the Processes Using Simulation

Dougbedji Fatondji, Andre Bationo, Ramadjita Tabo, James W. Jones, A. Adamou, and O. Hassane

Abstract In the drylands of Africa about 90% of the population is rural and depends on subsistence agriculture for their livelihoods. There is an increasing pressure on the natural resources due to the high population growth, and farmers are constrained to cultivate marginal lands, thereby compounding the land degradation problem. Low and erratic rainfall, its poor distribution within the growing season, prolonged dry spells, lack of adequate water supply due to soil physical degradation (soil crusting) and nutrient shortage adversely affect crop growth and yields. To address these problems, indigenous, easy to implement innovations such as the zai system may provide solutions to increase productivity. The effect of three planting techniques (Flat, zai pit of 25 cm and zai pit of 50 cm diameter) and three fertility management options (control, crop residue, cattle manure) were tested at Damari in 1999 in Niger. Soil water was monitored from weekly measurements using a Didcot

D. Fatondji (✉) • A. Adamou • O. Hassane
Soil & water management, International Crop Research Institute for the Semi-Arid Tropic,
P.O. Box 12404, Niamey, Niger
e-mail: d.fatondji@cgiar.org

A. Bationo
Soil Health Program, Alliance for a Green Revolution, in Africa (AGRA), Agostino Neto Road,
Airport Res. Area 6 Accra, PMB KIA 114, Airport-Accra, Ghana
e-mail: abationo@agra-alliance.org

R. Tabo
ICRISAT Niamey, P.O. Box 12404, Niamey, Niger

Forum for Agricultural Research in Africa (FARA), Accra, Ghana
e-mail: rtabo@fara-africa.org

J.W. Jones
Agricultural and Biological Engineering Department, University of Florida,
Gainesville, FL 32611, USA

J. Kihara et al. (eds.), *Improving Soil Fertility Recommendations in Africa using the Decision Support System for Agrotechnology Transfer (DSSAT)*,
DOI 10.1007/978-94-007-2960-5_6, © Springer Science+Business Media Dordrecht 2012

Wallingford neutron probe throughout the growing period. Data from that experiment were used to determine if the CERES-Millet model of the Decision Support System for Agrotechnology Transfer (DSSAT) is sufficiently robust to predict yield response to the zai water harvesting system. The model simulated the observed yield response of the control and the manure-amended plots with high r-square (0.99), low residual mean error square (340 kg·ha^{-1} for above ground biomass and 94 kg·ha^{-1} for grain yield) and high d-statistic (0.99), but this was not the case for the crop residue treatment, which was over-predicted. Soil water content and extractable soil water were also well simulated for the control and manure treatments. This evaluation of DSSAT provides a starting point for research to evaluate the performance of these technologies over wider areas in West Africa. The application of models for such studies must be interpreted in the context of limitations of the model to address some constraints. Nevertheless, the highly variable crop responses due to interacting effects of rainfall, management and adverse soil conditions in this region make this an extremely important approach in planning for technology adoption in an area and in interpreting results from experimental field research.

Keywords Zai • DSSAT • Simulation • Damari • Water haversting

Introduction

In the dry lands of Africa, about 90% of the population is rural and depends on subsistence agriculture for their livelihoods (Bationo et al. 2003). Low and erratic rainfall, its poor distribution within the crop growing period, prolonged dry spells, lack of adequate water supply due to soil physical degradation (soil crusting, low water retention) and nutrient shortage often adversely affect crop growth and yields in this zone (Zougmoré et al. 2003). According to Sundquist (2004) desertification along the Sahara desert proceeds at an estimated 1,000 km^2 every year, which further increases the pressure on arable land. One reason for this is the mounting population pressure (3% yearly growth on average) and the limited availability of fertile land. Many researchers have studied a wide range of management practices for increasing productivity, including testing of better adapted varieties, use of inorganic and organic fertilizer (Buerkert et al. 2002; Schlecht et al. 2004; Bationo et al. 1995; Yamoah et al. 2002; Tabo et al. 2007), rotation and residue management (Bado et al. 2007; Fatondji et al. 2006; Adamou et al. 2007) and water harvesting methods (Agyare et al. 2008; Roose et al. 1993).

One of the techniques studied is the zai system, an indigenous technology that combines rain water collection (Roose et al. 1993; Fatondji 2002), and nutrient management. Research has shown that the zai technology promotes crop production on highly degraded soils and helps alleviate the adverse effects of dry spells, which are frequent during the cropping period in the Sahel (Roose et al. 1993; Hassan 1996; Fatondji et al. 2006). This results not only from soil fertility improvement derived from the applied amendment and wind-driven materials that collect in the pits, but also improvement of the soil water status following the breakage of the surface crust and higher water infiltration (Fatondji 2002). Applying the zai technology on crusted soils results in rapid progress of the soil wetting front, which may drain to deeper layers to recharge ground water and also leach nutrients (e.g., nitrates Fatondji et al. 2011). Depending upon soil and crop growth conditions, the proportion of drained water is variable. Using the zai system or other soil and water conservation techniques for crop production may improve productivity and help eliminate hunger in the dry land of West Africa. However soil type, climate and other conditions vary over time and space and influence the ways those technologies interact.

Because many studies do not collect enough data to understand the interactive effects of soil and weather conditions that affect crop yield, it is difficult to extrapolate results from specific experiments to other soil and weather conditions. Crop simulation models deal with these interactive effects and have been used to predict how crop technologies will perform across sites and seasons and may help develop better management techniques for a wide range of conditions. However, it is not clear that the models are suitable for predicting crop performance under the degraded soils and extreme climatic conditions of West Africa. Although models have been used in many studies in Africa, they usually take into account only one or two limiting factors, such as variable rainfall and fertilizer input. Degraded soils have a number of factors that interact to limit crop growth and yield in complex ways. In order to use crop models for those conditions, they need to be tested in experiments in which measurements are made to provide all of the needed soil parameters, weather conditions, initial soil condition, management inputs and soil and crop growth responses. If the models are successful, they can be used to predict performance of the technologies and reduce the need for expensive and time-consuming field experimentation across regions.

Crop models in the Decision Support System for Agrotechnology Transfer (DSSAT) (Tsuji et al. 1994; Jones et al. 2003) have been used widely worldwide. This modeling system was designed for users to create computer experiments, simulate outcomes of the agricultural practices, soil, and weather conditions, and suggest appropriate solutions for specific sites (Jones et al. 1998). The millet model (CERES-Millet; Singh et al. 1991), like other models in DSSAT, is designed to be independent of location, season and management since it simulates the effects of weather, soil water, cultivar, and nitrogen dynamics in the soil on crop growth and yield. This model has not been evaluated for simulating production using zai technologies.

An experiment was conducted on a farmer's field at Damari in Niger (West Africa) to evaluate management systems that would increase yield and water productivity of millet (Fatondji et al. 2006). The overall objective of that work was to study the productivity and resource use efficiency of millet under rainfed conditions in the zai system as compared to flat planting on a highly degraded soil. In this study, we used data from that experiment to determine if the CERES-Millet model is sufficiently robust to predict yield response to the zai water harvesting system. This experiment was selected because of the potential importance of the zai system in the Sahel which has highly degraded soils and because an intensive set of data was collected on soil physical and chemical conditions, daily weather, weekly volumetric soil water versus depth, and crop yield and biomass productivities. The soil and climate conditions of this site challenge the capability of crop models because of the extreme soil physical and chemical properties and intensive rainstorms. The low soil water holding ability, soil crusting, low organic carbon, variable quality of organic amendments, low fertility, low pH and intensive rainfall events, when combined, may stretch the limits of crop models beyond their capabilities. In this study we hypothesized that millet crop performance and soil water status in the zai technology could be predicted with the CERES-Millet crop model using carefully measured weather and soil data at the experimental site. The specific objective was to evaluate the ability of this model to simulate the performance of millet in the zai system.

Material and Methods

Experimental Site

The experiment associated with the present study was conducted in 1999 in farmers' fields at Damari in Niger. Damari is located 45 km from Niamey, the capital city of Niger, at 13°12′ N and 2°14′ E. The long-term average annual rainfall is 550 mm, which falls between June and September. The long term monthly average minimum and maximum temperatures vary, respectively, between 16°C in January and 28°C in April and May and between 32°C in January and 42°C in April and May (Fig. 6.1). Monthly potential evapotranspiration (PET) is very high; monthly rainfall exceeds PET only in August (Sivakumar et al. 1993). During the experiment

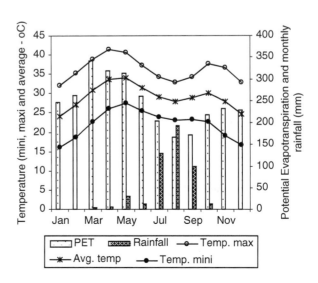

Fig. 6.1 Long term weather data at Damari; 1983–1999

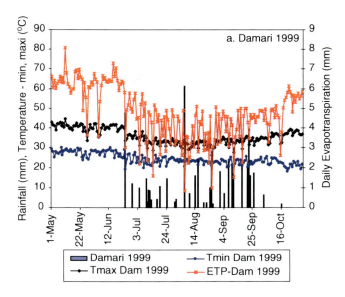

Fig. 6.2 Daily rainfall, minimum and maximum temperature, and potential evapotranspiration at the experiment site, Damari in 1999

in 1999, weather conditions followed this trend; total rainfall for the season was 499 mm (Fig. 6.2).

The soil at Damari is classified as kanhaplic Haplustult (American taxonomy–Soil Survey Staff 1998). It is acidic (pH-$H_2O = 3.6$–4.5), with 84% sand content, a relatively low effective cation exchange capacity (ECEC) (2.8 cmol kg^{-1}), and a very low soil water holding capacity (PAWC = 25–600 mm soil depth approximately). Because of intense rainfall events in the region the soils are prone to surface crusting (Casenave and Valentin 1989) and high runoff rates. The soil organic carbon ranged from 0.04% to 0.14% (Fatondji et al. 2006), even lower than the typical levels in Niger (about 0.22%, Bationo et al. 2003). The nutrient levels of soils in the region are very low (Bationo et al. 2003) and severely limit yield compared to the genetic potential of the rainfall environment. Table 6.1 shows measured chemical characteristics of the soil at the experiment site. Available P was far below the level of 2.1 mg/kg reported by Sinaj et al. (2001) typical to the soils of the Sahel, indicating the advanced degradation status of the soil. Total nitrogen was also very low compared to the average levels for Sub-Saharan Africa reported in Bationo et al. (1996). For other details on the experimental site, refer to Fatondji et al. (2006).

Despite these extreme conditions, farmers are forced to use them for producing crops because of limited land availability. Water harvesting technologies are therefore used to assure better soil water conditions to the crop. Due to the high price of mineral fertilizer, and also the risk of leaching nutrients, farmers use mostly organic

Table 6.1 Soil profile characteristics of the experiment field at Damari, measured in 1999*

Land management	Depth (cm)	pH (H$_2$O)	Total N (mg kg^{-1})	P-Bray 1 (mg kg^{-1})	C org (%)	Sand (%)	Clay (%)	Bulk density (g·cm^{-3})
Flat	15	3.9	0.13	1.73	0.14	84	13	1.6
	30	3.9	0.11	1.03	0.09	83	13	1.5
	45	3.7	0.12	0.74	0.07	84	13	1.5
	60	3.6	0.12	0.46	0.06	85	12	1.5
Zai 25 cm	15	4.6	0.11	1.03	0.09	84	13	1.5
	30	3.9	0.12	0.74	0.07	83	13	1.5
	45	3.7	0.12	0.46	0.06	84	13	1.5
	60	3.6	0.12	0.46	0.06	85	12	1.5
Zai 50 cm	15	4.2	0.11	1.03	0.09	84	13	1.5
	30	3.9	0.12	0.74	0.07	83	13	1.5
	45	3.7	0.12	0.46	0.06	84	13	1.5
	60	3.6	0.12	0.46	0.06	85	12	1.5

*Adapted from Fatondji et al. (2006)

manure in the zai technology. These amendments are often of variable quality with variable decomposition properties.

Field Experiment

The effect of planting technique (planting on flat vs. planting in zai pits of 25 cm diameter and pits of 50 cm diameter (15–20 cm deep) and amendment type (control, millet straw, and cattle manure) on millet growth and development were studied. The zai pits were dug in the dry season in the third week of May 1999. When digging the pits, the excavated soil was placed perpendicular to the slope on the lower side of the pit so that water flow would be oriented into the pit. The organic amendments were applied 36 days before sowing at the rate of 300 g dry weight per pit or pocket (identification of the planting hill on non-zai treated plots) (i.e. 3 t/ha for both manure and straw). When applied in the flat planting treatments, the amendment was incorporated to 5 cm depth to protect it from wind that could displace it. When applied in the zai pit, it was not initially covered, but it was covered later due to accumulation of sand and plant material blown and washed into the pit. The field was kept free of weeds throughout the growing season.

The millet straw used as amendment in the study had been collected from experimental fields at Sadoré and cut into small pieces of 10 cm length, whereas the cattle manure was collected from a barn on the same station. Urine was mixed with the feces, which increased N and K content and improved the quality. Table 6.2 presents the chemical composition of these amendments. The 2.53% N concentration of the

Table 6.2 Nutrient composition (%) of the organic material used in the experiment ad nutrient applied (kg ha^{-1})

Organic amendment	Nutrients content and C/N ratio				Nutrients applied per hectare (kg)		
	N (%)	P (%)	K (%)	C/N	N	P	K
Millet straw	1.18	0.10	1.57	50	32.7	2.8	43.5
Manure	2.53	0.94	1.72	21	62.9	23.3	42.8

Adapted from Fatondji et al. (2006)

manure was higher than the 1.2% N for cattle manure typically collected in farmers' corralled fields (Esse et al. 2001).

The experimental design was a Randomized Complete Block Design (RCBD) with two amendments + control (no organic amendment) and three planting techniques (nine treatments) replicated 4 times. Millet variety Sadore local was sown on 28 June at a planting density of 10,000 pockets per ha and harvested at maturity. Plants were thinned to three plants per pocket, approximately 3 weeks after planting.

Measurements

At maturity, grain and straw dry weight data were collected on a whole plot basis; one border row was left out on each side of the experimental units. The collected data were extrapolated to obtain yield on a per hectare basis.

Volumetric soil moisture contents were measured weekly at 15 cm intervals down to 210 cm depth using a neutron probe (Didcot Instrument Company Limited; Wallingford, UK). The probe had been calibrated *in-situ* for the soils of the experimental site applying the gravimetric method suggested by the manufacturer (Fatondji et al. 2006). The raw neutron probe data were converted to volumetric soil water contents (cm^3 cm^{-3}). Two 48 mm inner diameter aluminum access tube were installed in each experimental plot. One tube was installed between the planting pockets while the second was about 5 cm from the plants (in the pits in zai-treated plots). Data of the tubes installed close to the plant (on the pocket or in the zai pit) are reported in this chapter. The first measurements were made before the first rainfall on 7 June in 1999 and were continued throughout the growing period until harvest. To study the progress of the profile wetting, several dates were selected to match soil water measurements with other observations which were made in the experiment. These were the date of first measurement before planting, the day of planting as well as days of plant sampling. Only the top 60 cm soil water data were used in this study because roots did not extend below that depth. Extractable soil water was calculated as the difference between the volume of water at field capacity (or soil water drained upper limit, DUL) in the soil depth to a maximum rooting depth of 60 cm and the volume of water in the same soil profile at permanent wilting point (or lower limit). Rainwater productivity was calculated as ratio of aboveground biomass or grain yield to the amount rain between planting and grain harvest dates and was expressed in kg per millimeter of rain water. Drainage and runoff are important components of the water balance equation to calculate evapotranspiration. Runoff measurements were not taken during the experiment; therefore evapotranspiration water productivity values were not estimated.

CERES-Millet Model Simulation

The CERES-Millet model in DSSAT (Tsuji et al. 1994) was used to simulate the effects of the zai and amendment types on aboveground biomass production, grain yield, soil water content, and extractable soil water. Measurements of soil and weather conditions were used to provide the needed inputs to the model. These inputs are (1) the initial chemical and physical status of the soil, which was determined through soil characterization measurements made prior to the installation of the experiment on 12 May 1999. Soil samples were collected up to 200 cm depth to measure nutrient content and particle size distribution. (2) Nutrient content of the amendments were also measured. (3) Initial soil water conditions were determined using the first neutron probe measurement on 7 June before rain started. The CERES-based organic matter and nitrogen dynamics module (Godwin and Singh 1998) was used. (4) Weather data were collected with an automatic Campbell scientific weather station (daily rainfall, solar radiation, and minimum and maximum air temperature). (5) Lastly, data on the phenology of the crop were collected throughout the cropping season. For comparison with the simulated variables, actual crop yield and final biomass, soil water content, and extractable soil water data were obtained. The root mean square error (RMSE), mean absolute error (MAE) and d-statistic (Willmott 1981) were used to assess the agreement between simulated and observed values.

For the measurement and estimation of all soil parameters vs. depth in the zai treatments, the first measurement starts from the bottom of the pit as the rooting zone of the crop sown in the pit starts from this level. This was taken into account in the initial conditions and the soil characteristic input parameters for the model. Therefore 9 sets of initial conditions and soil analysis were used, which is equal to the number of treatment combinations tested. The fact that the first depth of soil water measurement started from the bottom of the zai pit did not have any influence on initial soil water content. In fact in the flat treatment, the first sampling layer started at the soil surface, whereas in the zai pit it started at the soil surface at the bottom of the pit.

Estimating Model Inputs Not Directly Measured

Some model parameters are difficult to measure directly and must be estimated from other measurements. In this study, three soil parameters were estimated using weekly soil water content vs. depth measurements in selected treatments: the lower limit of plant available water (LL), the drained up limit (DUL), and the surface

Table 6.3 Initial soil conditions set for the model from measured data in Damari in 1999

Land management	Depth (cm)	Initial soil water content (cm³·cm⁻³)	Soil water lower limits (LL) (cm³·cm⁻³)	Soil drained upper limit (DUL) (cm³·cm⁻³)	Saturation point (SAT) (cm³·cm⁻³)	Nitrate content (g[N]·Mg⁻¹ soil)
Flat	15	0.022	0.024	0.065	0.361	0.007
	30	0.028	0.024	0.075	0.354	0.004
	45	0.038	0.024	0.08	0.354	0.003
	60	0.042	0.024	0.08	0.358	0.002
Zai 25 cm	15	0.026	0.024	0.08	0.361	0.004
	30	0.037	0.024	0.08	0.354	0.003
	45	0.044	0.024	0.09	0.354	0.002
	60	0.046	0.024	0.09	0.358	0.002
Zai 50 cm	15	0.027	0.024	0.08	0.361	0.004
	30	0.038	0.024	0.08	0.354	0.003
	45	0.041	0.024	0.09	0.354	0.002
	60	0.046	0.024	0.09	0.358	0.002

water runoff curve number (ROCN). Although there are pedotransfer functions for estimating LL and DUL, from measured soil texture, these functions are not reliable for specific field sites (Gijsman et al. 2003). Genetic coefficients and an inherent soil productivity factor were estimated using maturity date, biomass and grain yield measurements in the manure treatments. Finally maximum root depth was estimated using soil pH and water measurements. Although these estimates were obtained by indirect methods, they are based on measurements that provided consistent predictions taking into account the many interacting factors.

To estimate the soil water LL, we took the average of the measured soil water contents of the first two soil layers (15 and 30 cm depths) taken on 7 June before the first rain of the season. Due to the long dry season from October, these first two layers were dry. The average volumetric water content of the two layers was 0.024 cm³·cm⁻³. We did not include the lower depths as higher values indicated that those layers probably did not reach the lower limit. To estimate DUL, the neutron probe readings were also used. Flat-planted control and 50 cm zai control treatments were used to estimate DUL for each layer for flat and zai treatments, respectively. The DUL was set to approximate the soil water values measured after rainfall had wet the soil, but before plants started rapidly extracting water. The results are reported in Table 6.3

Due to soil crusting, runoff was high in flat-planted treatments. Therefore a high coefficient (ROCN) was set for this treatment by comparing the time series of measured and simulated soil water contents in the control and manure flat-planted treatments. Iteratively, ROCN values were changed until simulated soil water vs. depth and time of season in these two treatments were in good agreement with observed soil water contents. Following this procedure, a runoff coefficient of 98.4 was obtained for the flat treatment. To estimate the runoff coefficient for the two zai pit

Table 6.4 Genetic coefficients for the millet variety used in the study

Parameter	Initial values (variety CIVT)	Values
Thermal time from seedling emergence to the end of the juvenile Phase (P1)	180	170
Critical photoperiod or the longest day length (in hours) at which development occurs at a maximum rate (P20)	12	12
Extent to which phasic development leading to panicle initiation (expressed in degree days) is delayed for each hour increase in photoperiod above (P2R)	150	150
Thermal time (degree days above a base temperature of 10°C) from beginning of grain filling (3–4 days after flowering) to physiological maturity(P2OP5)	500	450
Scalar for relative leaf size (G1)	2	1
Scalar for partitioning of assimilates to the panicle (G5)	0.50	0.77
Phylochron interval; the interval in thermal time (degree days) between successive leaf tip appearances. (PHINT)	43	43

sizes, we calculated the proportion of area occupied by a zai relative to the total area per pocket (1 m²). Although water falling between the pits has a chance to be captured in the pits, for simplicity we assumed that any drop falling between the pits would runoff at a rate determined by the ROCN of the flat treatment and all rain falling on the area of the pit would be retained. Based on this assumption we calculated a weighted average ROCN using the runoff coefficient of the flat planting and relative area of the zai hole to the area not in the hole. Therefore we obtained a ROCN of 93.5 for the zai of 25 cm diameter and 79.1 for the zai of 50 cm diameter. One ROCN was used for each planting technique regardless of amendment.

Genetic Coefficients

Genetic coefficients were estimated using measured biomass and grain yield and physiological maturity date for the zai manure treatments. Ideally, genetic coefficients should be estimated using data collected in experiments without water and nutrient stresses, but this is not possible in many cases such as in this experiment. Following Boote et al. (2003) and Mavromatis et al. (2001), coefficients for a variety in the DSSAT millet cultivar file was initially used, and phenology coefficients (P1, P2R, and P2OP5) were adjusted so that the simulated maturity date closely approximated the mean observed date for the manure treatments (good, least nutrient stress treatment) (Table 6.4.). Afterward, coefficients that determine biomass production and its partitioning into grain yield were considered. However, this was done simultaneously with adjustments to the soil fertility factor (SLPF), which must be used to account for limited nutrients in the soil that are not included in the model. Other researchers (e.g., Singh et al. 1994; Naab et al. 2004) found that SLPF values ranging between 0.63 and 1.00 were necessary for some soils in India and Ghana.

In this case, it was noted that soil P levels were very low, which justified our modification of this factor. Thus, G5, the parameter that partitions assimilates into grain, and SLPF were modified together using both grain and biomass yield as criteria.

Maximum rooting depth is determined by a root growth factor (SRGF) in each soil layer. Layers down to the maximum root depth have values computed from the DSSAT software, and values below that were set to 0.0. In this study, it was assumed that due to Al toxicity and low pH below 30 cm depth, roots would not grow below 30 cm. This was consistent with the neutron probe data that showed no soil water extraction below that depth. Therefore SRGF was set to zero for all layers below that depth.

Results and Discussion

Experiments

Soil water content. Figure 6.3 shows graphs of soil water content vs. depth for different measurement dates for the 25 cm zai and flat planted treatments. The same trend was observed for both pit sizes, but soil water contents were higher for the plots with the 50 cm diameter zai. The wetting front was already below 200 cm on the day of planting in the zai treated plots (Fig. 6.3a, c, e), whereas in the non-zai treated plots, it was shallower on the same date (Fig. 6.3b, d, f). The results indicate that even though the structure of the soil is sandy, breaking the surface crust and digging the pits was highly favorable for water infiltration compared to the flat treatment. Volumetric soil water content (VWC) was still higher at deeper layers in the zai vs. the flat treatments even towards the end of the season. In the control-zai plots for instance, at 200 cm depth, VWC was about 0.08 cm^3/cm^3 compared to 0.051 cm^3/cm^3 for the flat-control treatment.

In general, the soil water profile was shallower in the manure treated plots than the other treatments regardless of the type of planting technique. Towards the end of the cropping season, in the zai as well as on flat treatment with cattle manure, soil water content decreased significantly compared to plots treated with millet straw, indicating high water consumption of the crop due to increased biomass production. Particularly in flat treatment amended with cattle manure, the wetting front remained at 60 cm during the whole growing period, which is an indication that the presence of crust hampered water infiltration. But in addition crop water uptake may have increased considerably due increased biomass production

Extractable soil water (ESW) was calculated based on a maximum depth of 60 cm for comparison with the output of the model. In the flat-planted plots, extractable soil water was lower than in the zai-treated plots regardless of the amendment type (Fig. 6.4). This was more pronounced in the manure treated plot probably due to higher plant consumption as reported in Fatondji et al. (2006). Biomass and grain yield on these plots were high compared to the control. The same thing may have happened in the 25 cm diameter zai amended with manure where extractable soil

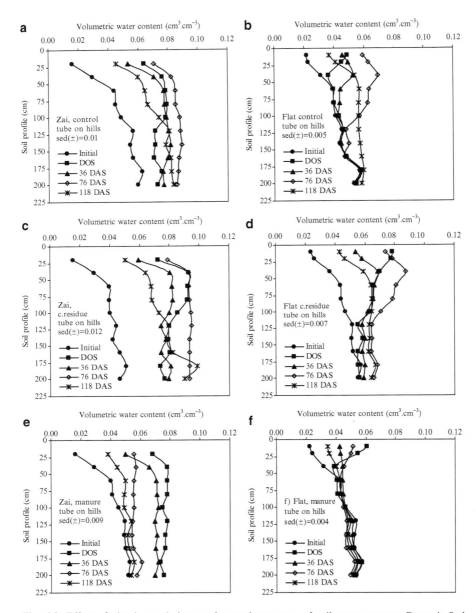

Fig. 6.3 Effect of planting technique and amendment type of soil water content. Damari; *Sed* standard error of difference between means

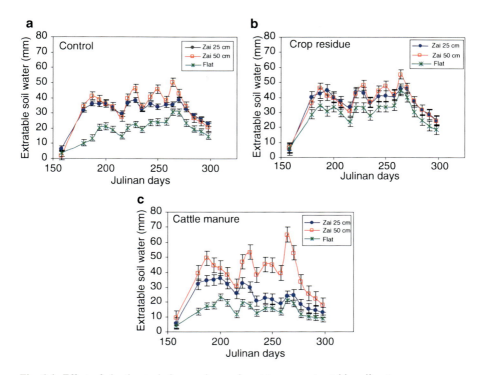

Fig. 6.4 Effect of planting technique and amendment type on extractable soil water

water dropped substantially from day 240 until the end of the cropping season, which was not the case of the zai of 50 cm diameter. This is an indication that more water was collected in pits of 50 cm diameter.

<u>Yield components.</u> In the control non-amended plots, the zai treatment increased above ground biomass yield by a factor 3 for both pit sizes, while grain yield increased by a factor 19 for pit size 25 cm diameter and 9 for pit size 50 cm (Table 6.5). Nevertheless the yields were extremely low compared to the average millet grain yield in Niger, which is 300 kg ha^{-1} (Bationo et al. 1989). No farmer would crop a field that would produce 1 kg ha^{-1} of grain. This is an indication that crop production would not be possible without external nutrient inputs in the soil where the experiment was conducted. It also shows that water is not the major constraint. Only minor yield increases were observed in the zai compared to flat planting when crop residue was applied.

The zai treatment significantly increased the above ground biomass (5,133 kg ha^{-1} and 5,711 kg ha^{-1} for the 25 cm and 50 cm zai, respectively) compared to 2,967 kg ha^{-1} for flat planting, when cattle manure was applied. Grain yield also increased (1,156 kg ha^{-1} and 1,100 kg ha^{-1} for 25 cm zai and 50 cm zai, respectively) compared to 705 kg ha^{-1} for flat planting (Table 6.5). This shows that by breaking the crust with the zai digging, better conditions were created for crop growth. This may

D. Fatondji et al.

Table 6.5 Millet above ground biomass and grain yield as affected by planting technique under various fertility management conditions, Damari 1999

| Treatments | Yield data (kg ha^{-1}) | |
	Above ground biomass	Grain
Control		
Flat planting	96	1
Zai 25 cm	303	17
Zai 50 cm	280	8
Sed ($\pm$)	90.4	4.9
F.prob	ns	0.045
Crop residue		
Flat planting	795	127
Zai 25 cm	1,059	168
Zai 50 cm	924	157
Sed ($\pm$)	232.5	48.9
F.prob	ns	ns
Manure		
Flat planting	2,967	705
Zai 25 cm	5,133	1,157
Zai 50 cm	5,711	1,100
Sed ($\pm$)[a]	872.8	276.1
F.prob	0.044	ns

Adapted from Farondji et al. (2006)

[a]*Sed* Standard error of difference between means

have also helped the crops to escape from the effect of dry spells. In the Sahel, and particularly during this experiment, dry spells resulting in 2 weeks without rain were frequent (Fatondji et al. 2006). The observed differences were statistically significant only for millet aboveground biomass yield when cattle manure was applied and for grain yield of the control non-amended plots. No statistically significant differences were observed between the zai pit sizes in terms of above ground biomass and grain yield. This could be due to high variability in the data because of the harsh conditions of the experiment, particularly in the control and the crop residue amended plots. The residual mean square error was even higher than the treatment mean for crop residue amended plots. Nevertheless, we speculate that soil nutrient content was so low, that water availability alone without application of nutrients made only small differences in crop productivity among the soil management techniques.

In flat-planted plots, 2,900 kg ha^{-1} of aboveground biomass yield was obtained with manure application compared to 1,200 kg ha^{-1} for crop residue and 96 kg ha^{-1} for the control non-amended plot. Relatively high grain yield production was also obtained with manure application in flat-planted plots (705 kg ha^{-1}) compared to 127 kg and 1 kg ha^{-1} for crop residues and control non-amended plots, respectively (Table 6.6). All observed differences were statistically significant. In both zai pit sizes, manure application significantly increased aboveground biomass and grain yield compared to the crop residue and the control treatments. In the 25 cm diameter

Table 6.6 Millet above ground biomass and grain yield as affected by amendment type under various soil management conditions, Damari 1999

Treatments	Above ground biomass (kg ha^{-1})	Grain yield (kg ha^{-1})
Flat planting		
Control	96	1
Crop residue	1,159	127
Manure	2,967	705
Sed(±)	597.3	215.4
Fprob	0.008	0.036
Zai 25 cm		
Control	303	17
Crop residue	1,195	1,68
Manure	5,133	1,157
Sed(±)	666.2	112.6
Fprob	<0.001	<0.001
Zai 50 cm		
Control	280	8
Crop residue	924	157
Manure	5,711	1,100
Sed(±)	406.9	79.8
Fprob	<0.001	<0.001

Sed Standard error of difference between means
Adapted from Fatondji et al. (2006)

zai treatment amended with manure, aboveground yield increased by a factor 17 compared to the control and a factor of 4 compared to crop residue treatments. The same trend was observed in grain yield but more pronounced as manure application in the 50 cm zai treatment increased yield by factors of 138 and 7 for the control and crop residue treatments, respectively. Grain yield in crop residue-amended plots was higher than the control by a factor 4 and 7 in the 50 and 25 cm zai treatments, respectively. All the differences were highly significant statistically (Table 6.6). These results show that for better results with the zai technology, there is a need of additional nutrient input of good quality. Nevertheless, due to the excess water that would collect in the zai, it may be preferable to use organic amendment for nutrient input instead of inorganic fertilizers that tend to leach with water drainage.

Table 6.7 shows the effects of organic amendment type on observed rain water productivity and simulated results for the same parameter for comparison. Manure application in the zai resulted in above ground yield of 12 kg·mm^{-1} of rain on average versus 0.6 kg mm^{-1} of rain for the control treatment. Grain yield per mm of rain water also increased by a factor 64 and 128 for zai 25 cm and zai 50 cm, respectively, compared to the control non-amended plots. On flat-planted plots, manure application increased rain water productivity by a factor 31. All the observed differences were statistically significant. When compared to flat planted

Table 6.7 Rainfall water productivity as affected by amendment type under various soil management practices; Damari 1999

| | Rain water productivity (kg·mm^{-1}) | | | |
| | Above ground yield | | Grain | |
	Observed	Simulated	Observed	Simulated
Zai 25 cm				
Control	0.67	0	0.04	0
C.residue	2.65	6.6	0.37	1.82
Manure	11.38	13.4	2.56	3.11
Sed(±)	1.476		0.249	
Fprob	< 0.001		< 0.001	
Zai 50 cm				
Control	0.62	0	0.02	0.00
C.residue	2.05	6	0.35	1.62
Manure	12.66	12.9	2.44	2.82
Sed(±)	0.902		0.18	
Fprob	< 0.001		< 0.001	
Flat				
Control	0.21	0	0	0
C.residue	2.57	5.7	0.28	1.42
Manure	6.58	6.2	1.56	1.52
Sed(±)	1.324		0.477	
Fprob	0.008		0.036	

C.residue Crop residue, *Sed* Standard error of difference between means

plots, manure application in the zai improved aboveground biomass and grain yield per mm of rain by a factor 2. These results indicate that the crop made better use of rain water in the zai when manure was applied. Similar results were reported by Fatondji et al. (2006) on another experimental site where the same technologies were tested.

Simulated Results

Soil water content. Figure 6.5 shows the simulated soil water contents for the control flat planted treatment for soil layers 5–15 and 15–30 cm compared to the observed data. In general there is a good prediction of the movement of the wetting front in the 5–15 cm layer during all the sampling period except for the 4th and the 14th sampling dates, which correspond to period of successive rainfall events (the first was taken 1 day after 3 days of rain (total of 26 mm) and the second was taken 1 day after a rain event of 21 mm) which were under-predicted. The seventh and tenth samplings, which were taken after 11 and 7 days of dry spells, respectively, were over-predicted. In general the model performed poorly in predicting soil water

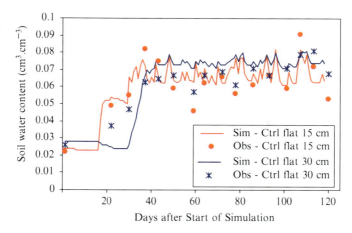

Fig. 6.5 Soil water content in the flat control treatment, soil layers 5–15 and 15–30 cm. Simulated (*Sim*) and observed (*Obs*) data

content in the 15–30 cm layer. It grossly under-predicted the second sampling, which was 4 days after 48 mm of rain and the third sampling, which was 2 days after 10 mm of rain. Subsequently, the model predicted high water content all along the sampling period except for samplings 11, 13 and 14, which were taken 4 days after cumulative rains of 39, 38 and 20 mm and which were all well predicted. In both soil layers, the dry spells were over-predicted. Our observation is that there is no consistent trend on which we could base our argument in relation to the prediction of soil water content depending upon the number of days before or after rainfall event. Nevertheless we have to admit that the time resolution of this model would not allow this level of detail.

Overall, the ability of the model to simulate over most of the season was good as supported by the low residual mean square error (RMSE = 0.01), high d-statistic of 0.9 and r-square of 0.7. Figure 6.6 shows the simulation results of the control zai for the same soil layers. The general trend was well simulated, although the model simulated more peaks that were not observed from the field measurements. We also point out that the model can be off by one day, since it is not indicated exactly when during a day the rainfall occurred, and at what time measurements were taken. Figure 6.7 shows the simulated soil water contents in the manure amended zai 50 cm plot. In general the measured trend was captured; nevertheless the 11th sampling that was after 4 days of cumulative rainfall of 39 mm was not simulated accurately by the model. Actually, there was no consistent relationship between the trends in observed rainfall events and the time of sampling as per our observations. Nevertheless, further studies may be needed to address these details which may help us understand why the model over-predicted water content in some cases while in other less water content was predicted compared to the observed values.

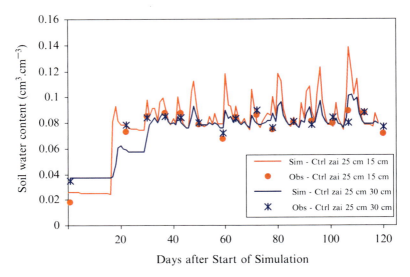

Fig. 6.6 Soil water content in the zai control treatment, soil layers 5–15 and 15–30 cm. Simulated (*Sim*) and observed (*Obs*) data

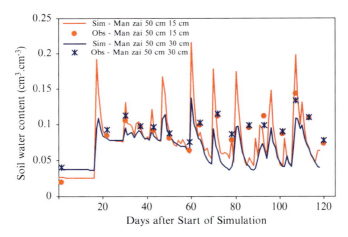

Fig. 6.7 Soil water content in the zai 50 cm manure treatment, soil layers 5–15 and 15–30 cm. Simulated (*Sim*) and observed (*Obs*) data

Figure 6.8 shows the extractable soil water in the top 60 cm of soil. The observed trend was captured by the model in the control flat and manure zai 50 cm with high d-statistics (0.913 and 0.821) except for the samplings 11 and 12, which were over-estimated as observed already with the graph of soil water content in the 50 cm zai treatment amended with manure. The manure flat treatment had a lower d-statistic (0.618) and very low r-square of 0.37; but a fairly low RMSE (5.953). One of the major inputs of the zai technology is the breakage of the soil crust while digging the

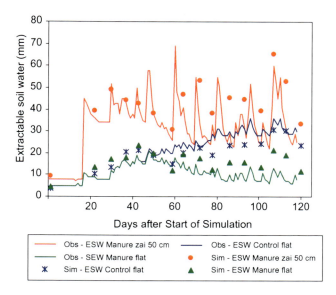

Fig. 6.8 Extractable soil water (*ESW*) in the control and manure flat + manure zai 50 cm treatments; Simulated (*Sim*) and observed (*Obs*) data

zai holes. Depending upon the size of the pit, the amount of water collected may differ. The results presented here were obtained by using a different runoff curve number for each land management treatment (flat, zai 25 cm and zai 50 cm). The general trend was that almost no water extraction occurred below the top 45 cm of soil, which according to Fatondji (2002) was the depth above which 98% of the plant roots (dry weight basis) were concentrated. High Al content (29%) of the experiment soil (Fatondji et al. 2006) hampered root growth beyond the zone of application of the organic amendment, which could explain why water extraction did not occur at those depths. Nevertheless some water would move upward from the 30–60 cm layer as the upper layer dries out, and thus plants will extract some of the water from the 30–60 depth due to diffusion even if roots are not in that layer.

Yield Components

Table 6.4 shows the genetic coefficients estimated for the Sadore local variety used in this experiment. The estimated value of SLPF was 0.68. These genetic coefficients and the SLPF were then used for all other treatments in the experiment. Table 6.8 shows statistics comparing simulated vs. observed biomass and grain yield for flat-planted and zai treatments with manure and with no amendments (six treatments). Although the manure treatments were used to estimate these parameters, these results demonstrate a good ability of the millet model to simulate differences among these six treatments, with high r-square and d-statistics and low root

Table 6.8 Simulated vs. observed aboveground biomass and grain yield – statistical comparisons

Variable name	Mean (kg·ha⁻¹)			Mean (kg·ha⁻¹)		RMSE	
	Observed	Simulated	r-Square	Difference	Abs.Diff.	(kg·ha⁻¹)	d.stat
Control, crop residue and manure							
Total biomass	1,974	2,441	0.847	467	711	943.2	0.948
Grain yield	382	591	0.728	209	225	327.755	0.885
Control and manure only							
Total biomass	2,415	2,339	0.988	−76	290	340.026	0.995
Grain yield	498	536	0.991	39	63	93.045	0.993

d-stat d-statistic Willmott (1981), *Abs.Diff.* Absolute difference

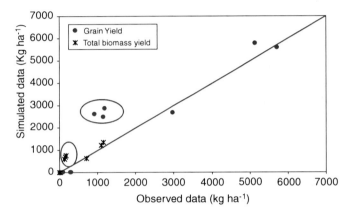

Fig. 6.9 Observed versus simulated top dry matter and grain yield. *Circled points* are for the crop residue treatments

mean square errors between simulated and observed data. Due to the marginal conditions of the experiment in terms of soil physical and chemical characteristics, the crops were so stressed in some treatments that they could hardly grow and simulated growth was very low relative to the crop's genetic potential (control non-amended plots). Although these conditions are extreme to be simulated by conventional models, these results show that by setting the parameters for conditions in this experiment the model was able to simulate the observed responses to these six treatments.

However, the simulation results for the crop residue treatments substantially over-predicted observed yields (Fig. 6.9 – circled symbols). The model may have under-predicted immobilization of N following addition of the high C:N residues in this treatment.). One other possibility is that the low response observed in the experiment relative to simulated yield may have been due to the very low phosphorus content in the crop residue. Phosphorus concentration in the crop residue was 0.10%, whereas it was 0.94% for cattle manure. The version of the millet model in DSSAT v4.5 used in this study did not account for phosphorus limitations to growth, although this option is available for other crops (Dzotsi 2007). This means that simulated yields for this experiment were based only on water and nitrogen availability in addition to weather and genetic coefficients. When soils have very low phosphorus levels, and very little or no phosphorus is applied as an amendment or fertilizer, the model may over-predict biomass growth and grain yield, which is what happened in the crop residue treatments. It is also possible that the nutrient content of the crop residue was highly variable and inputs for this amendment were not accurate. In manure treatments, as manure decomposed it released about 9 times more phosphorus than the decomposing crop residue, which apparently favored crop growth and yield. This trend is not captured by the model as the phosphorus module is not yet available in DSSAT for millet.

The model performed very well in terms of rain water productivity, particularly for manure and control plots; whereas the effect of crop residue was not well captured in the model outputs (Table 6.7). This implies that water productivity estimated by the model can be used to estimate rain water productivity for comparing the zai with manure amendment vs. control flat planted management systems in other years or locations if the required soil, weather, amendment, and planting technique model inputs are known.

Discussion and Conclusions

This study addressed the challenge of simulating low productivity of millet due to combination of crusting soils (with adverse effects on soil water balance), extremely low PAWC soils, N + P constraints on crop growth and low and erratic rainfall. We explored the possibility of simulating millet production in one of the extreme conditions that farmers have to deal with using an experiment in which detailed data were collected on soil physical and chemical properties, organic amendment properties, weather, yield components, and weekly soil water content vs. depth measurements for nine treatment combinations of planting techniques and organic amendments. These carefully-collected data provided a good test of how well the millet model would predict the range of responses that were observed. But even with the extensive data set, we found that several input parameters needed by the model had to be estimated using indirect methods. Although this need may exist in other conditions, the model was highly sensitive to these uncertain inputs for the conditions at this site. The most sensitive inputs that had to be estimated indirectly were the genetic coefficients for the variety used in the experiment, the runoff curve number for different zai vs. flat planted treatments, and the soil fertility factor. Nevertheless, simulated yield results were very good for the manure treatments on flat planted and zai treatments, predicting aboveground biomass values that ranged between about 3,000 and 5,700 kg ha^{-1} and grain yield ranging between about 700 and 1,200 kg ha^{-1}. For control treatments, simulated aboveground biomass and grain yield values were below 10 kg ha^{-1}, whereas observed values were somewhat higher. But since observed grain yields were less than 20 kg ha^{-1}, these treatments all represented crop failure.

The output of the model in terms of rain water productivity as a ratio of dry matter or grain production to the amount of rain from planting to harvest was captured when compared to the observed data particularly for manure treated and control plots. This indicates that the measured data were adequately used to estimate the model parameters. The model calculates the amount of rain received during the cropping period based on weather data provided. Therefore this result could be expected as those two treatments were well simulated in terms of aboveground biomass and grain yield.

Water harvesting techniques are one of the means to combat desertification in sub-Saharan Africa. They are mostly used on the prevailing degraded bare land of the region. The results of the field study that served as the basis for model evaluation demonstrated that the zai technology is a powerful tool, which under extreme physical and chemical conditions, can substantially increase crop yield and provide conditions for crops to escape from adverse effects of dry spells. Even though zai technologies are indigenous in some countries, there is a need to extend them for broader use. A study for evaluating their effectiveness across environments is therefore needed because among the water harvesting technologies, the zai is simple and easy to implement by farmers as it requires locally-available material.

We contend that simulation analysis of these options can be used to provide insight on the effectiveness of alternative management systems. However, realistic inputs are needed for environments to be studied, and results must be interpreted relative to uncertainties in the inputs as well as limitations in the models. For example, the comparison of manure amendments in zai vs. flat planting, based on these results, could be simulated for a range of similar soils and climates with a reasonably high confidence level. However, simulating the use of lower quality amendments, particularly in similar highly degraded soils, would need to be interpreted in the context of limitations of the model. Although this is always true for model applications, the harsh conditions in this region make this an extremely important issue when conducting and interpreting results from such studies.

References

Adamou A, Bationo A, Tabo R Koala S (2007) Improving soil fertility through the use of organic and inorganic plant nutrient and crop rotation in Niger. In: Bationo A, Waswa B, Kihara J, Kimetu J, (eds) Advance in integrated soil fertility management in sub-Saharan Africa: challenges and opportunities. Proceeding of AfNet international symposium, Yaounde Cameroun, 17–21 Sept 2004. Springer, Dordrecht, pp 589–598

Agyare WA, Antwi BO, Quansah C (2008) Soil and water conservation in Ghana: practices research and future direction. In: Bationo A, Tabo R, Waswa B, Okeyo J, Kihara J, Fosu M, Kabore S (eds) Synthesis of soil water and nutrient management research in the Volta Basin. Ecomedia Ltd Publisher, Nairobi

Bado BV, Bationo A, Lompo F, Cescas MP, Sedogo MP (2007) Mineral fertilizers, organic amendments and crop rotation managements for soil fertility maintenance in the Guinean zone of Burkina Faso (West Africa). In: Bationo A, Waswa B, Kihara J, Kimetu J (eds) Advance in integrated soil fertility management in sub-Saharan Africa: challenges and opportunities.

Proceedings of AfNet international symposium, Yaounde Cameroun, 17–21 Sept 2004. Springer, Dordrecht, pp 589–598

Bationo A, Christianson CB, Mokwunye AU (1989) Soil fertility management of pearl millet-producing sandy soil of Sahelian west Africa: the Niger experience. In: ICRISAT International Crop Research Institute for the Semi-Arid Tropics 1989. Soil, crop and water management systems for rainfed agriculture in the Sudano-Sahelian zone. Proceedings of an international workshop, 7–11 Jan 1987. ICRISAT Sahelian Center, Niamey, Niger

Bationo A, Sedogo MP, Buerkert A, Ayuk E (1995) Recent achievements on agronomic evaluation of phosphorus fertilizer sources and management in the west Africa semi-arid tropics. In: Ganry F, Campbell B (eds) Sustainable land management and African semi-arid and sub-humid region. Proceeding of the SCOPE workshop, Dakar, Senegal, 15–19 Nov 1993. CIRAD, Montpellier, pp 99–109

Bationo A, Rhodes E, Smaling EMA, Visker C (1996) Technologies for restoring soil fertility. In: Mokwunye AU, de Jager A, Smaling EMA (eds) Restoring and maintaining the productivity level of West African soils: key to sustainable development. IFDC-Africa, LEI-DLO abd SC-DLO, Miscellaneous Fertilizer Studies No. 14, International Fertilizer Development Center, Muscle Shoals (USA)

Bationo A, Mokwunye U, Vlek PLG, Koala S, Shapiro BI (2003) Soil fertility management for sustainable land use in the West African Sudano-Sahelian zone. In: Gichuru MP et al (eds) Soil fertility management in Africa: a regional perspective. Academy Science Publisher & Tropical Soil Biology and Fertility, Nairobi, pp 253–292

Boote KJ, Jones JW, Bactchelor WD, Mafziger ED, Myers O (2003) Genetic coefficients in the CROPGRO-soybean model: links to field performance and genomics. Agron J 95:32–51

Buerkert A, Piepho HP, Bationo A (2002) Multi-site time trend analysis of soil fertility management effect on crop production in sub-Saharan West Africa. Exp Agric 38:163–183

Casenave A et, Valentin C (1989) Les états de surface de la zone sahelienne; Influence sur l'infiltration. Les processus et les facteurs de réorganisarion superficielle. (ed) ORSTOM – Institut Français de Recherche Scientifique pour le Développement en Coopération. Collection Didactiques, Paris 1989, pp 65–190

Dzotsi KA (2007) Comparison of measured and simulated responses of maize to phosphorus levels in Ghana. MS thesis, Agricultural and Biological Engineering Department, University of Florida, Gainesville, 175 pp

Esse PC, Buerkert A, Hiernaux P, Assa A (2001) Decomposition and nutrient release from ruminant manure on acid sandy soils in the Sahelian zone of Niger, West Africa. Agr Ecosyst Environ 83:55–63

Fatondji D (2002) Organic amendment decomposition, nutrient release and nutrient uptake by millet (Pennisetum glaucum) in a traditional land rehabilitation technique (zaï) in the Sahel. PhD Thesis, Ecological and Development Series No 1. Center for development research, University of Bonn, Cuvillier Verlag, Gottingen

Fatondji D, Martius C, Bielders C, Vlek P, Bationo A, Gérard B (2006) Effect of planting technique and amendment type on pearl millet yield, nutrient uptake, and water use on degraded land in Niger. Nutr Cycl Agroecosyst 76:203–217

Fatondji D, Martius C, Bielders C, Vlek P, Bationo A (2011) Effect of zai soil and water conservation technique on water balance and the fate of nitrate from organic amendments applied: A case of degraded crusted soils in Niger. In A. Bationo et al. (eds) Innovations as key to the green revolution in Africa, 1115 DOI 10.1007/978-90-481-2543-2_114

Gijsman AJ, Jagtap SS, Jones JW (2003) Wading through a swamp of complete confusion: how to choose a method for estimating soil water retention parameters for crop models. Eur J Agron 18:77–106

Godwin DC, Singh U (1998) Nitrogen balance and crop response to nitrogen in upland and lowland cropping systems. In: Tsuji GY, Hoogenboom G, Thornton PK (eds) Systems approaches for sustainable agricultural development; understanding options for agricultural production. Kluwer Academic, Boston, pp 55–77

Hassan A (1996) Improved traditional planting pits in the Tahoua department, Niger. An example of rapid adoption by farmers. In: Chris R (ed) Sustaining the soil. Indigenous soil and water conservation in Africa. Earthscan, London, pp 56–61

Jones JW, Tsuji GY, Hoogenboom G, Hunt LA, Thornton PK, Wilkens P, Imamura DT, Bowen WT, Singh U (1998) Decision support system for agrotechnology transfer: DSSAT v3. In: Tsuji GY, Hoogenboom G, Thornton PK (eds) Systems approaches for sustainable agricultural development; understanding options for agricultural production. Kluwer Academic, Boston, pp 157–177

Jones JW, Hoogenboom G, Porter CH, Boote KJ, Batchelor WD, Hunt LA, Wilkens PW, Singh U, Gijsman AJ, Ritchie T (2003) The DSSAT cropping system model. Eur J Agron 18(3–4): 235–265

Mavromatis TK, Boote KJ, Jones JW, Irmak A, Shinde D, Hoogenboom G (2001) Developing genetic coefficients for crop simulation models with data from crop performance trials. Crop Sci 41:40–51

Naab JB, Singh P, Boote KJ, Jones JW, Marfo KO (2004) Using the CROPGRO-peanut model to quantify yield gaps of peanut in the Guinean savanna zone of Ghana. Agron J 96:1231–1242

Roose E, Kabore V, Guenat C (1993) Le"zaï": Fonctionnement, limites et amélioration d'une pratique traditionnelle africaine de réhabilitation de la végétation et de la productivité des terres dégradées en région soudano-sahelienne (Burkina Faso). - Cahier de l'ORSTOM, Serie Pedologie XXVIII (2):159–173

Schlecht E, Hiernaux P, Achard F, Turner MD (2004) Livestock related nutrient budgets within village territories in western Niger. Nutr Cycl Agroecosys 70:303–319

Sinaj S, Buerkert A, El-Hadjj G, Bationo A, Traore H, Frossard E (2001) Effect of fertility management strategies on phosphorus bioavailability in four West African soils. Plant Soil 233:71–83

Singh U, Ritchie JT, Thornton PK (1991) CERES-Cereal model for wheat, maize sorghum, barley and pearl millet. Agron Abstract 78

Singh P, Boote KJ, Yogeswara Rao A, Iruthayaraj MR, Sheikh AM, Hundal SS, Narang RS, Singh P (1994) Evaluation of the groundnut model PNUTGRO for crop response to water availability, sowing dates, and seasons. Field Crops Res 39:147–162

Sivakumar MVK, Maidukia A, Stern RD (1993) Agroclimatology of West Africa: Niger, 2nd edn. Information Bulletin 5. ICRISAT, Patancheru, 116 pp

Soil Survey Staff (1998) Keys to soil taxonomy, 8th edn. USDA/NRCS, Washington, DC

Sundquist B (2004) Land area data and aquatic area data; a compilation, 1st edn. March, 2004. http:/home.Alltel.net/bsundquist1/la0.html [checked 5.12.2005]

Tabo R, Bationo A, Gerard B, Ndjeunga J, Marchal D, Amadou B, Garba MA, Sogodogo D, Taonda JBS, Hassane O, Diallo MK, Koala S (2007) Improving cereal productivity and farmers' income using a strategic application of fertilizers in west Africa. In: Bationo A, Waswa B, Kihara J, Kimetu J (eds) Advance in integrated soil fertility management in sub-Saharan Africa: challenges and opportunities. Proceeding of AfNet international symposium, Yaounde Cameroun, 17–21 Sept 2004. Springer, Dordrecht, pp 589–598

Tsuji GY, Uehara G, Balas S (1994) Decision support system for agrotechnology transfer (DSSAT) v3. International Benchmark Sites Network for Agrotechnology Transfer, University of Hawaii, Honolulu

Willmott CJ (1981) On the validation of models. Phys Geogr 2:184–194

Yamoah CF, Bationo A, Shapiro B, Koala S (2002) Trend and stability analysis of millet yield treated with fertilizer and crop residue in the Sahel. Field Crop Res 75:53–62

Zougmoré R, Kambou NF, Zida Z (2003) Role of nutrient amendments in the success of half-moon soil and water conservation practice in semiarid Burkina Faso. Soil Till Res 71:143–149

Chapter 7
Effect of Integrated Soil Fertility Management Technologies on the Performance of Millet in Niger: Understanding the Processes Using Simulation

A. Adamou, Ramadjita Tabo, Dougbedji Fatondji, O. Hassane, Andre Bationo, and T. Adam

Résumé La faible fertilité des sols et la rareté des pluies sont les facteurs les plus limitatifs de la production agricole dans la zone soudano-Sahélienne en Afrique de l'Ouest. La région habite les populations les plus pauvres de la planète dont 90% vivent en milieu rural et tirent leur nourriture d'une agriculture de subsistance. Cependant, les rendements des céréales en général et du mil en particulier qui constituent la nourriture de base sont très faibles (300–400 kg/ha). La recherche a développé des technologies de gestion intégrée de la fertilité des sols mais elles n'ont pas été adoptées par les paysans. DSSAT (Decision Support System for Agrotechnology Transfer) est un outil incorporant des modèles de 16 différents types de cultures avec un logiciel facilitant l'évaluation et l'application des modèles de cultures pour différentes utilisations. Mais son utilisation requiert un minimum de données sur le climat, les sols, les cultures et aussi les données expérimentales. Les simulations obtenues à partir de ces données permettront aux chercheurs de développer beaucoup de résultats prometteurs en milieu paysan. Cette étude montre

A. Adamou (✉)
TSBF, ICRISAT Niamey, P.O. Box 12404, Niamey, Niger
e-mail: a.abdou@icrisatne.ne

R. Tabo
Deputy Director, Forum for Agricultural Research in Africa (FARA), Accra, Ghana

D. Fatondji
GT/AE, ICRISAT Niamey, P.O. Box 12404, Niamey, Niger

O. Hassane
GIS, ICRISAT Niamey, P.O. Box 12404, Niamey, Niger

A. Bationo
Senior Resource Mobilization, Africa, Alliance for a Green Revolution
in Africa (AGRA), Accra, Ghana

T. Adam
Faculté d'Agronomie, Université Abdou Moumouni, Niamey-Niger

J. Kihara et al. (eds.), *Improving Soil Fertility Recommendations in Africa using the Decision Support System for Agrotechnology Transfer (DSSAT)*,
DOI 10.1007/978-94-007-2960-5_7, © Springer Science+Business Media Dordrecht 2012

les interactions entre la fertilité des sols et les rendements de mil dans trois sites (Banizoumbou, Bengou et Karabedji) au Niger sur une périodes de 5 ans (2001–2005) et une simulation dans DSSAT sur l'azote.

Abstract Low soil fertility and erratic rainfall are the most limiting factors to crop production, in the Sudano-Sahelian zone of West Africa. The region is the home of the world's poorest people, 90% of whom live in villages and gain their livelihood from subsistence agriculture. However, yields of cereals in general, and millet in particular that constitute the staple food of rural people, are very low (300–400 kg/ha). Research has developed technologies of integrated soil fertility management, but resource poor farmers have not adopted them. DSSAT (Decision Support System for Agrotechnology Transfer) is a tool incorporating models of 16 different crops with software that facilitates the evaluation and application of crop models for different purposes. Its use requires a minimum data set on weather, soil, crop management and experimental data. The simulations from these data can help scientists to develop promising management options to improve farmer's conditions. However, requirements for such model use is to evaluate its capabilities under farming situations, soils, and weather that are characteristic of the area where it will be used. This study was conducted to evaluate the DSSAT millet model capabilities for simulating the interactions between soil fertility and millet yields in three sites (Banizoumbou, Bengou and Karabedji) of Niger over 5 years (2001–2005) and different nitrogen management.

Introduction

Soil degradation, loss of organic matter, low soil fertility and yields, poverty, and climatic changes are among the main factors reducing crop production in the world. In the Sahel, low rainfall and its variability and distribution, dry spells and other climatic factors affect crop production. Production losses are mainly due to drought (2/3) and cricket attack (1/3) (Nanga 2005). Water balance in the region is positive only during 3 months of the year; meaning that water is still a limiting factor for crop production in a region where 90% of the population is rural and depend on subsidence rainfall for agriculture. Millet is the main crop in Permanent Inter-State Committee for Drought Control in the Sahel (CILSS) countries with 45% of cereal production followed by sorghum (28%) and maize (11%). Niger is second after Burkina Faso with 27% of cereal production in CILSS. Niger, with a population of 12.94 million in 2006, is one of the food-deficit countries in the world (CILSS/Agrhymet 2005). Only 12% of the country has an annual rainfall of 600 mm or more and only 10% has 350–600 mm. Cereal crop needs at least 300 mm if it is well distributed (Moustapha 2003). Cereal production in 2004/2005 was estimated to about 2.50 million tons with a negative balance of about 0.22 million t, which is equivalent to 7.5% of Niger population needs (Nanga 2005). Niger is the poorest country in the world according to the United Nations Development Program (UNDP) classification based on Human Development Index (HDI). Production was less than demand in Niger in two of the last 5 years. Ninety-eight percent of the cereal production in Niger is from rainwater. Rice, the principal irrigated crop, is less than 2% of the total cereal production. Water availability for irrigation also depends on rainfall, and in 2004 rice production decreased to only 0.5% of cereal production in Niger (FAO 2004). As Sahelian agriculture depends on rainfall, which varies considerably from year to year with considerably effects on crop production. Poor rainfall, high temporal variability and spatial distributions, and other climatic constraints are characteristic of the Sahel. This makes water a principal constraint of crop production in this country.

Population pressure has reduced cultivable area and traditional fallow is no longer feasible in Niger. It's known that millet is a crop adapted to Sahelian climate conditions, but combined with low soil fertility, low rainfall can greatly reduce crop productivity. In this context of soil degradation and poor climatic conditions, recommended farmer's practices are not appropriately adapted. ICRISAT research aims to find and propose to farmers combinations of soil fertility technologies, methods to increase water use efficiency and varieties to significantly improve crops yields.

This study was conducted to help identify improved natural resources management in these poor soils and weather conditions. Phosphorus, nitrogen, manure and rainfall effects on crop production are shown in the study. DSSAT v4.02 (Jones et al. 2003; Hoogenboom et al. 2004) was used to compare simulated and measured data and specific effects of climatic and fertility factors on millet production.

Materials and Methods

Three sites were used for this study due to agronomic and climatic data availability: Banizoumbou, Bengou and Karabédji, where ICRISAT has conducted experiments and where meteorological stations were used to record daily weather data that are needed for the millet model. We used data from 5 years (2001–2005) on trials conducted in the three different sites. Mineral and organic fertilizers were used and comparisons between the control and other treatments show the low productivity of the farmer's system. Among limiting factors, water was included and simulations were done to show its effects on millet production. Many factors are used in DSSAT but only climate and fertility will be used in this study.

Before the rainy season there is always uncertainty of what to be planted, when and how. Climatic data such as rainfall, minimum and maximum temperatures, and solar radiation are necessary inputs to the model. Other data are also used as DSSAT inputs: soil parameters, fertilizer type, manure and other organic fertilizers from the three sites.

There is no significant difference between soils of Karabédji and Banizoumbou but annual rainfall vary from 300 to 500 mm. At Bengou, rainfall is about 800 mm per year and the soils are very different. This large variation in rainfall may result in

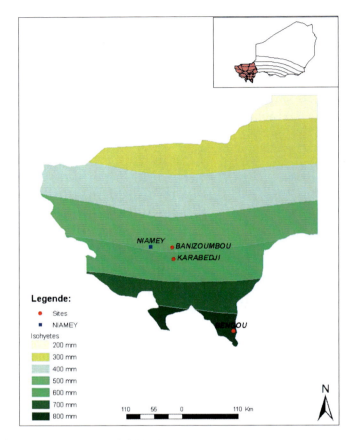

Fig. 7.1   Study sites and annual rainfall isohyets

significant differences in millet production in Niger for production situations where water is the most limiting factor.

Study Sites

Climate

These three sites are in a normal rainfall range for millet production (above 300 mm) but Bengou is more humid with more than 700 mm per year and over 4–5 months of rainfall compared to the two other sites where it rains only 3 months per year. Annual rainfall of Banizoumbou and Karabédji are, respectively, 360 and 450 mm (Fig. 7.1).

Table 7.1 shows that rainfall during the past 5 years exceeded the longterm mean is mentioned but not provided anywhere. At Banizoumbou and Karabédji, only

Table 7.1 Annual rainfall (mm), 2001–2005

Sites	2001	2002	2003	2004	2005	Mean
Banizoumbou	344	345	510	363	444	*401*
Karabédji	368	638	557	524	468	*511*
Bengou	985	732	884	630	784	*803*

Table 7.2 Soils characteristics at the different sites

Sites	pH KCl	C.org (%)	P-Bray1 (mg/kg)	Ca^{2+} Cmol/kg	ECEC Cmol/kg	N_{min} (mg/kg)
Banizoumbou	4.4	0.12	1.5	0.4	0.8	5
Karabédji	4.2	0.16	1.9	0.2	0.8	4
Bengou	4.2	0.33	2.5	0.4	1.3	9

2001 rainfall was under its long-term mean. At Bengou 2 years (2002 and 2004) received less than the normal amounts.

Soils

Soils from Banizoumbou and Karabédji are similar sols ferrugineux tropicaux lessivés. With more than 90% sand, these soils have low organic matter, low ECEC; and are very poor in nutrient contents. Soils from Bengou are slightly better and have higher organic carbon (more than 0.2%) and higher ECEC (Table 7.2).

Fields close to the village receiving high organic matter due to human and animal activities are more productive compared to the outfields. Prudencio (1993) observed such fertility gradients between fields closest to the homestead (home gardens/infields) and those furthest (bush fields/outfields). Soil organic carbon contents of between 11 and 22 g/kg have been observed in home gardens compared with 2–5 g/kg soil in bush fields. Fofana et al. (2006) in a comparative study at Karabédji-Niger on degraded lands (bush fields) and non degraded (infields) have observed that millet grain yield across years and fertilizer levels averaged only 800 kg/ha in bush fields and 1,360 kg/ha on infields. Recovery of fertilizer N applied varied considerably and ranged from 17% to 23% on bush fields and from 34% to 37% on infields. Similarly, recovery of fertilizer P was 18% for bush fields and 31% for infields over 3 years of cropping. It is clear that degraded soils are poor in organic carbon, their responses to fertilizer applications are less, and the recovery of fertilizer applied is very low. Soil degradation was defined by FAO (2002) as the loss of soil productivity capacity in term of decreased fertility, biodiversity and natural resources. Yield loss due to soil degradation in Africa varied from 2% to 50% the last 10 years (Scherr 1999). Bationo et al. (2006a) in Sherr (1992) and in Oldeman et al. (1992) in a description of degradation of arable soils in Africa and in the rest of the world estimated that degraded soil proportions were 38% in the world and 65% in Africa. During the last 30 years, nutrient losses in African soils were equivalent to 1,400 kg/ha N (urea), 375 kg/ha of SSP (phosphorus) and 896 kg/ha of KCl (potassium). In

Niger, Henao and Baanante (2006) estimated nutrient losses of 56 kg/ha (NPK) during the 2002–2004 cropping seasons.

Crops

At Karabédji and Banizoumbou, crops are mainly millet, sometimes intercropped with cowpea. At Bengou, millet is intercropped with sorghum, groundnut or cowpea as it rains up to 5 months per year.

Long-term average millet and sorghum grain yields are, respectively, 400 and 190 kg/ha. In 2002 and 2003, respective cereal production was 3.34 and 3.56 million tons in Niger for millet and sorghum grain yields were respectively 461 and 476 kg/ha (FAO 2004). Research at ICRISAT (1985) showed that in the semi-arid zones of the Sahel where annual rainfall is over 300 mm, nutrients are more limiting than water in crop production. At Sadore (Niger), with 560 mm of annual rainfall, 1.24 kg of millet grain per millimeter of water was harvested without fertilizer and 4.14 kg of millet grain per millimeter of water when fertilizer was used (Bationo et al. 2006).

Experimental Layout

Three sites where ICRISAT conducted studies and where climatic and physical conditions are different were used in this study: Banizoumbou, Bengou and Karabédji. The selected Experiments started in 2001 and are still on going. They are factorial experiments with $3 \times 3 \times 3 = 27$ treatments on 4 replications with 3 levels of phosphorus (0, 13 and 26 kg P/ha), 3 levels of manure (0, 2 and 4 t/ha) and 3 levels of nitrogen (0, 30 and 60 kg N/ha). Climatic data were collected over years: rainfall by using rain gauges at each site and temperature and solar radiation collected from a nearby meteorological station.

The trial was established for calculating the fertilizer equivalency of manure and to compare mineral and organic fertilizers use efficiencies. Manure nutrient composition was analyzed every year and used as inputs in DSSAT. Grain and total dry matter yields were measured in this study and used to compare with simulated data. Because DSSAT v4.02 did not have a phosphorus model, we used GENSTAT to analyze P responses and DSSAT for analyzing the climatic effect on productivity in this study. Effect of fertilizers, sites and years on millet production can be analyzed and climatic effect can be shown through grain and total dry matter yield comparisons.

Initial conditions are characterized by soil analysis for N, P, and K and were used in DSSAT. Simulations were conducted to show the role of water, fertilizers (mineral and organic) and other climatic factors. Water and fertilizer limitations were estimated to compare the results in different cases and their impact in millet production.

Nitrogen effects are included in DSSAT v4.02 and were used to study this nutrient's effects on millet production. Measured and simulated data were compared and differences were interpreted. Growth stages and water and nitrogen stresses were simulated in DSSAT to compare their respective limitations. We also compared the different rainfall amounts in each growth stage to evaluate the effects of water stress at different stages. In this paper, only the Banizoumbou site was used for the DSSAT simulations, but the other three sites results were also used to analyze and interpret results .

Results and Discussions

Effect of Fertilizers in Millet Production

Overall analysis of data from the three sites over the 2001–2005 years showed that phosphorus was more significant than the other nutrients in millet grain production. P alone accounted for 17% in the total variation followed by manure (8%) and nitrogen (4%). These trends of phosphorus, manure and nitrogen were the same for millet stover production (26%, 13% and 8%, respectively) The factor year accounting for 29% of the total grain plus stover yield variation decreased to only 4% for millet grain production and the site factor accounted for 2% in both stover and grain production. This high variation show that stover production varied less than grain production due to grain losses from bird attacks, which reduced 2001 grain production at Bengou. Other factors such as water stress at grain filling can also affect grain yield. Water stress during the grain filling period is highly important for grain yield and should be highlighted in the simulated results.

Soil fertility levels should be increased as nothing can be done to increase rainfall, but rural populations are very poor and fertilizer costs are very high. In Oumou and Ed Heinemann (2006), Africa accounted for only 3% of the world fertilizer consumption with 13% of world's arable soils and 12% of world's population. Sub-Saharan Africa (excluding south Africa) accounted for less than 1% in the world fertilizer consumption, equivalent to 9 kg/ha compared to an amount of 148 kg/ha in Asia and the Pacific region. In 2002, fertilizer consumption in Niger was only 1.1 kg/ha where 1 t of fertilizer cost $400 compared to $90 in Europe whereas average income in Niger was less than $1 per day making fertilizer unaffordable.

Phosphorus

Millet grain production over the 5 years increased with phosphorus rates, but variations were higher per year, especially in 2003 when yields were higher at all P rates (Fig. 7.2b). Millet stover yields also increased with phosphorus; increases were about the same across years (Fig. 7.2a). If fertilizer were affordable for

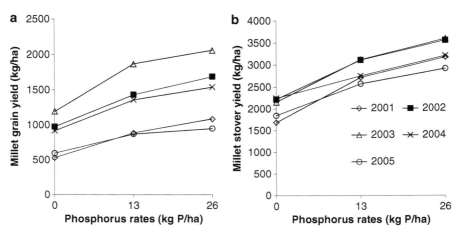

Fig. 7.2 Effect of phosphorus on millet grain (**a**) and stover (**b**) yields

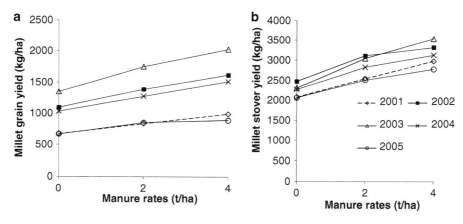

Fig. 7.3 Effect of manure on millet grain (**a**) and stover (**b**) yields

farmers, the hill placement of small quantity (4 kg P/ha) can double millet grain yield. Tabo et al. (2006) showed that a micro-dose of 4 kg P/ha increased millet and sorghum grain yields up to 43–120% and farmer's income were improved by 52–134% in the studied countries (Burkina Faso, Mali and Niger).

Manure

Similar to phosphorus, manure use showed that grain yields varied over years and increased with high manure application rates. In 2003, grain yield was higher but stover yield remained about the same as in other years (Fig. 7.3).

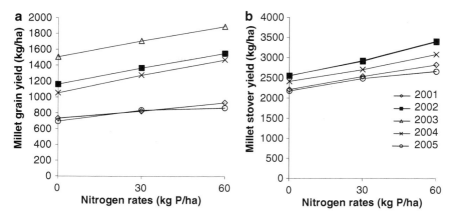

Fig. 7.4   Effect of nitrogen on millet grain (**a**) and stover (**b**) yields

Nitrogen

Nitrogen treatments resulted in the same variability on grain yield as phosphorus and manure and the same trends on millet stover production (Fig. 7.4). The best grain yields were observed in 2003.

Effect of Sites on Millet Production

Overall, the site factor accounted for only 1.6% of the total variation. The same trends were observed in the grain production with 17.8% and 4% of the variation, respectively, for the phosphorus and manure treatments. But the year factor became highly significant with 29% of the total variation and decreased to only 4% for the millet stover while other factors (P, manure and N) accounted for 26%, 13% and 8% of the variation, respectively. Millet stover production was about the same over the years, meaning that only grain yields show a significant year-year variation.

Fertilizers and Sites

Phosphorus

Phosphorus effects on grain production was more important at Bengou than at the others sites where there were similar grain production levels (Fig. 7.5). Stover production was reversed at Bengou, which registered the lowest yield although the annual rainfall was higher.

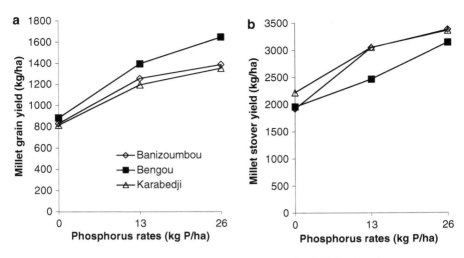

Fig. 7.5 Effect of phosphorus on millet grain (**a**) and Stover (**b**) yields in three sites

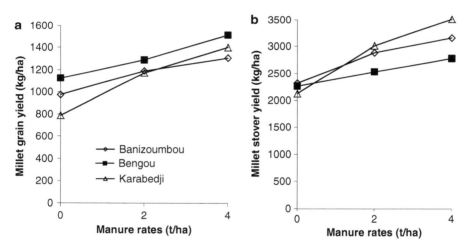

Fig. 7.6 Effect of manure on millet grain (**a**) and stover (**b**) yields in three sites

Manure

Manure effect was the same under the three sites with only a small increase in millet grain production at Bengou (Fig. 7.6).

Nitrogen

Compared to phosphorus and manure effects, the nitrogen effect was lower at Bengou but showed the same effects as the other sites on millet grain production. This effect is reversed in stover production. Fertilizers showed that their effectiveness was higher

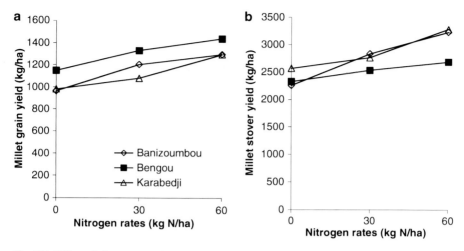

Fig. 7.7 Effect of nitrogen on millet grain (**a**) and stover (**b**) yields in three sites

on grain production at Bengou than at the other sites where the nutrients seemed to be used during vegetative plant growth to produce more biomass (Fig. 7.7).

Water and Fertilizer Use Efficiencies

Water Use Efficiency (WUE)

Water use efficiency (WUE) was increased with fertilizer use. The mean over 5 years was only 0.6 kg of grain per millimeter of water for the control and 2.1–2.6 kg of grain per millimeter of water when fertilizer was used. WUE in 2003 was the best over the 5 years, demonstrating a good correlation between WUE and yields. WUE is higher at Bengou where it increases to 3.6 kg of grain per millimeter of water.

Fertilizer Use Efficiency (FUE)

Compared to WUE, FUE was higher with low rates of fertilizer and the trends were the same at the 3 sites. PUE was 2–3 times higher than NUE (nitrogen use efficiency), confirming that phosphorus was the most limiting factor in millet production at the studied sites. FUE was also higher in 2003 and poorer in 2005 for both phosphorus and nitrogen.

Correlation Between WUE, FUE and Yields

A good correlation of WUE and grain yield was observed with different fertilizer sources ($r^2 = 1$): high WUE gives high yields. But there is no good correlation between FUE and grain yield (Fig. 7.8).

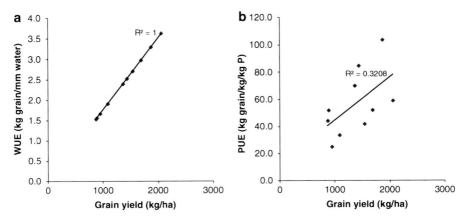

Fig. 7.8 Water (**a**) and phosphorus (**b**) use efficiency

Other Site Factors

Since fertilizer effects vary over the different sites, each one should have its own characteristic such as initial soil conditions. Bengou had higher responses to fertilizer; but no correlation was found to prove that this was due to the high rainfall specific to that site. On all other sites, high yield was not a consequence of high rainfall, but good rainfall distribution did result in high yield in 2003 for all three sites. In addition, Bengou showed higher soil fertility values than the two other sites as was shown in Table 7.2.

Year Effects on Millet Production

As expected, millet yields varied from year to year (Fig. 7.9). Important factors for these include rainfall, temperatures, solar radiation, wind, and crop pests and diseases. The first three factors are used in DSSAT and should be more characteristic of a particular year than the others. The agrometeorological variables are the main data for crop simulation model: minimum and maximum temperatures, solar radiation and total rainfall (Hoogenboom 2000). The other factors can also contribute to yield levels, such as bird attacks that occurred in 2001 at Bengou. Wind, a source of soil degradation and erosion, can also affect crop productivity.

In this sites, millet grain yield varied over years while stover yields were stable. Yields were high in 2003 but annual rainfall was not (except for Banizoumbou). It is clear that rainfall contributes to millet production, but it is not the only important factor affecting millet production.

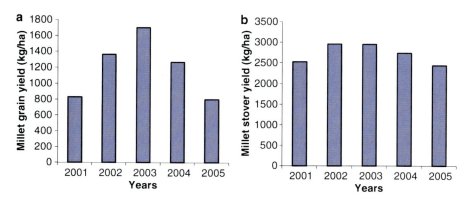

Fig. 7.9 Annual millet grain (**a**) and stover (**b**) yields observed in three sites

Over years, phosphorus, manure and nitrogen accounted for 45%, 22% and 11% of the total variation of millet grain production, respectively. They accounted respectively for 26%, 13% and 8% of the total variability while the site factor accounted for only 4% and 2% for grain and stover production, respectively.

Factors Other Than Rainfall

Temperature

Minimum and maximum temperatures of Katanga used for Karabédji and Banizoumbou followed the same trends over the 5 years. Temperatures were high from March to May, low from June to August, increased in September to October, and decreased again from November to February. In 2003, temperatures varied from the general trends with the lowest temperatures among all years except from July to September when temperatures are generally low but were high in 2003. In 2005 temperatures were in general higher than normal. Grain production was high in 2003 and low in 2005, showing a correlation between temperatures and grain production.

At Bengou, minimum temperatures were higher than the other sites but followed the same trends with long periods of low temperatures during the rainy season. High yields were observed during 2003 when temperatures were low and low yields during 2005 when temperatures were high.

Solar Radiation

No particular effect was observed with solar radiation, but it decreased at Gaya particularly from April to September corresponding to the low temperature period.

In general solar radiation affects panicle initiation duration as shown by Alagarswamy et al. (1998) in the experiment on sorghum varieties where young plants become photoperiod sensitive. When solar radiation is high, panicle initiation occurs more rapidly. But there was not much variation of solar radiation among sites or years in this study. Also, there was no significant variation in panicle initiation among the sites, years, and treatments in this study.

Other Rainfall Factors

Mostly, high rainfall is needed for rainfed crops to produce high yields. But millet yield may not be highly correlated with total seasonal rainfall due to the importance of other factors and due to the timing of rainfall events and related drought periods during a season. For example, at Banizoumbou, from planting to harvest, only 269 mm of rain was received in 2001 but millet yield was higher in 2001 than in 2005 when 353 mm was received. In 2005, more dry spells occurred than in 2001; timing of rain is very important in millet production.

Annual Rainfall

Annual rainfall and crop yield are not well correlated. If so, Banizoumbou, Karabédji and Bengou yields should have been highest in 2002, 2002 and 2001, respectively. However, for the three sites, 2003 had the highest yield even though rainfall that year was not the highest.

Rainy Season Length

When rainfall and monthly rain frequency and season length were analyzed, it was clear that high variations in grain yield were due to dry spells during the grain-filling period. Local varieties with long cycles of 110–120 days started their grain filling around 70–80 days after sowing, and any dry spell after this time can affect negatively millet grain production. ICRISAT research on millet varieties comparing grain filling in drought conditions has shown lower grain weights by 22% in 1988 and 27% in 1989. Treatments under water stress have shown a short grain filling period compared to irrigated ones.

At Banizoumbou and Karabedji, some dry spells occurred during the grain filling period (10–13 days) and during the whole cycle. In 2003, the longest dry spell was only 5 days during the grain filling period and grain yield was high compared to the other years. At Bengou, a long dry spell occurred in 2005, from 58 days after sowing (DAS) until harvest, and grain yield was considerably decreased.

Soil fertility affects millet production more than climatic factors at these sites, although it varies over years. 2003 and 2005 have shown some particular maximum and minimum yields due to some particular low and high temperatures, affecting water balance during the crop cycle.

Growth Development and Water Stress

Water and nitrogen stress analyses showed that farmers' practice (no fertilizer or manure applications) was not affected by water stress, but that nitrogen stress started at 20 DAS to harvest. It's clear that water was not the most limiting factor for farmers' practice, but that fertilizer was in this study.

In West Africa, drought risks are more related to the mean annual rainfall. With increasing annual rainfall, the percentage frequencies of short dry spells increases while the frequencies of long dry spells decreases. In general, dry spells around panicle initiation are higher than those during the flowering phase, particularly for locations with low rainfall. The dry spells become progressively longer at some point during the grain-filling phase. At low-rainfall locations, this occurs much earlier than at locations with higher rainfall (Sivakumar 1991).

Water stress was particularly high for the 60 kg/ha N treatment in 2001, 2004 and 2005. Annual rainfall for these years showed durations of 98, 75 and 107 days, respectively. The shortest rainfall duration was observed in 2004 but the lowest yield was in 2005, meaning that high temperatures in 2005 also affected water balance and yields.

Yields and Plant Growth Simulations

Simulations showed biomass development and grain formation during the growth cycle. Simulated yields were higher than observed ones except for farmers' practice (N0) where the simulated yields were sometimes lower (Tables 7.3 and 7.4). The higher simulated yields may be explained by the inability of the model to simulate phosphorus since the millet model in DSSAT did not have a P component. Phosphorus is an important nutrient in western Africa and Bationo et al. (2006) showed that nitrogen and manure applications are more efficient when combined with P.

Conclusion

There is always risk in rainfall crop production. In millet, crops under fertilizer treatments are more affected by water stress as shown by DSSAT simulations. During the 5 years of experiments, crops were never destroyed by water stress and yield was increased by the use of fertilizer in all cases. The farmer's practice was not

Table 7.3 Grain yields simulated (*S*) and measured (*M*) of treatments N0, N30 and N60 (Banizoumbou, 2001–2005)

Year	2001		2002		2003		2004		2005	
Trts	S	M	S	M	S	M	S	M	S	M
N0	104	290	353	339	264	365	201	354	433	328
N30	1,145	520	1,401	750	1,256	500	710	615	1,285	542
N60	1,300	635	2,522	542	2,372	584	965	908	1,996	526

Table 7.4 Stover yields simulated (*S*) and measured (*M*) of treatments N30 and N60 (Banizoumbou, 2001–2005)

Year	2001		2002		2003		2004		2005	
Trts	S	M	S	M	S	M	S	M	S	M
N0	293	1,275	1,671	1,238	1,147	1,458	858	1,523	1,862	1,523
N30	4,421	2,040	8,202	2827	7,376	1,937	5,313	2,289	6575	2,289
N60	6,046	2,280	13,025	2,867	12,266	2,575	8,585	2,508	9,397	2,508

Trts treatments

affected by water stress caused by dry spells because of the more limiting effects on nutrients. This means that annual rainfall was enough for millet crops under farmer's practices, and fertilizer (especially P) is more limiting since crop yields under fertilizer were improved even though some dry spells occurred during the cropping season. It was also shown that only water stress during the grain filling period affected millet grain yield and stover production was not affected by dry periods during grain filling. Farmer's practice was always deficient in nutrients with and without water stress.

Water stress affecting crop yield occurred in September corresponding to the end of the rainy season. Although we cannot control weather, we can adapt our cropping systems to get the maximum benefit from rainfall. If millet production is low in Niger, it is mainly because high producing technologies adapted to the weather conditions are not adopted by farmers. Simulations on nitrogen have showed that simulated yields were always better than yields under farmer's practice, indicating that there is potential to increase millet yields by adding N fertilizer. However, the simulated responses to N were higher than measured, mainly due to the fact that P is a major limiting factor in these studies and the DSSAT model did not include this factor. This highlights the fact that there is a need to incorporate a soil P model to address the major nutrient limitation in these soils.

If water is limiting to crop production, others factors also contribute in the Sahelian context. Although annual rainfall was sometime higher in the studied sites, yields were still lower. DSSAT is a tool that can be used to determine which factors most affect crop production, especially N fertilizer and water. Simulated and measured data can be compared to select high productive systems. It would be also good to extend the study area to other agro-climatic zones.

References

Alagarswamy G, Reddy DM, Swaminathan G (1998) Duration of the photoperiod-sensitive phases of time to panicle initiation in sorghum. Field Crop Res 55:1–10

Bationo A, Hartemink A, Lungu O, Naimi M, Okoth P, Smailing E, Thiombiano L (2006a) Africa soils: their productivity and profitability of fertilizer use; Background paper prepared for the African Fertilizer Summit, Abuja, Nigeria, 9–13 June 2006

Bationo A, Waswa B, Kihara J, Kimetu J (eds) (2006b) Advantages in integrated soil fertility management in Sub-Saharan Africa: challenges and opportunities. Springer, Dordrecht. Nutr Cycl Agroecosys 76, 2–3 Nov 2006

CILSS/AGRHYMET : Rapport final: Réunion de concertation technique sur les perspectives des productions et bilan céréalier ex-post 2004/2005 et prévisionnel 2005/2006 des pays du CILSS : Nouakchott du 7 au 10 novembre 2005

FAO (2002) Land degradation assessment in drylands. FAO, Rome, 18 pp

FAO (2004) Rapport spécial au Niger, 21 décembre 2004

Fofana B, Woperies MCS, Bationo A, Breman H, Tamelokpo A, Gnakpenou D, Ezui K, Zida Z, Mando A (2006) Millet nutrient use efficiency as affected by inherent soil fertility in the West African Sahel (in press)

Henao J, Baanante C (2006) Agricultural production and soil nutrient mining in Africa: implications for resource conservation and policy development. International Fertilizer Development Center (IFDC), Muscle Shoals

Hoogenboom G (2000) Contribution of agrometeorology to the simulation of crop production and its application. Agr Forest Meteorol 103:137–157

Hoogenboom G, Jones JW, Wilkens PW, Porter CH, Batchelor WD, Hunt LA, Boote KJ, Singh U, Uryasev O, Bowen WT, Gijsman AJ, du Toit A, White JW, Tsuji GY (2004) Decision support system for agrotechnology transfer version 4.0 [CD-ROM]. University of Hawaii, Honolulu

ICRISAT (International Crop Research Institute for the Semi-Arid Tropics) (1985) Annual report 1984. ICRISAT Sahelian Center, Niamey

Jones JW, Hoogenboom G, Porter CH, Boote KJ, Batchelor WD, Hunt LA, Wilkens PW, Singh U, Gijsman AJ, Ritchie JT (2003) DSSAT cropping system model. Eur J Agron 18:235–265

Moustapha A (2003) Les céréales au Niger : de la production à la commercialisation; dans: forum1. inter-reseaux.net article 371

NangaJ (2005) Grain de Sable.Com n°15 – mai 2005: Mali & Niger : la mondialisation néolibérale contre les plus pauvres

Oumou Camara, Ed Heinemann (2006) Overview of the fertilizer situation in Africa; Africa Fertilizer Summit, , Abuja, Nigeria, 9–13 June 2006

Prudencio CY (1993) Ring management of soils and crops in west African semi-arid tropics: the case of the Mossi farming system in Burkina Faso. Agric Ecosyst Environ 47:237–264

Scherr SJ (1999) Past and present effects of soil degradation. In: Scherr SJ (ed) Soil degradation-a threat to developing-country food security by 2020. International Food Policy Research Institute, Washington, DC, pp 13–30. 2020 Discussion paper 27

Sivakumar MVK (1991) Durée et fréquence des périodes sèches en Afrique de l'Ouest : Bulletin de recherche n°13. ICRISAT Patancheru, Andhra Pradesh 502324, Inde

Tabo R, Bationo A, Gérard B, Ndjeunga J, Marchal D, Amadou B, Annou MG, Sogodogo D, Taonda JBS, Hassane O, Diallo MK, Koala S (2006) Improving cereal productivity and farmers' income using a strategic application of fertilizers in West Africa. In: Bationo A, Waswa BS, Kihara J, Kimetu J (eds) Advances in integrated soil fertility management in Sub-Saharan Africa: challenges and opportunities. Springer, Dordrecht

Chapter 8
Evaluation of the CSM-CROPGRO-Soybean Model for Dual-Purpose Soyabean in Kenya

A. Nyambane, D. Mugendi, G. Olukoye, V. Wasike, P. Tittonell, and B. Vanlauwe

Abstract Limited information is available on the potential performance of introduced dual purpose varieties across different Kenyan soils and agro-ecological environments and consistency across sites and seasons. Crop simulation modeling offers an opportunity to explore the potential of and select introduced cultivars for new areas before establishing costly and time-consuming field trials. Dual purpose soybeans were introduced due to their ability to improve soils and at the same time provide substantial grain yields. The objective of this study was to derive genetic coefficients of recently introduced dual purpose soybean varieties and to explore the reliability of the Cropping System Model (CSM)-CROPGRO-Soybean model in

A. Nyambane (✉)
Department of Environmental Sciences, Kenyatta University,
P.O. Box 43844-00100, Nairobi, Kenya

Tropical Soil Biology and Fertility Institute of the International Centre
for Tropical Agriculture (TSBF-CIAT), P.O. 30677, Nairobi, Kenya
e-mail: alfnbs@yahoo.com

D. Mugendi • G. Olukoye
Department of Environmental Sciences, Kenyatta University, P.O. Box 43844-00100,
Nairobi, Kenya
e-mail: dmugendi@yahoo.com

V. Wasike
Kenya Agricultural Research Institute (KARI), P.O. Box 57811, Nairobi, Kenya
e-mail: vwwasike@yahoo.com

P. Tittonel
CIRAD, Mount pleasant Harare, Zimbabwe
e-mail: Pablo.tittonell@cirad.fr

B. Vanlauwe
Tropical Soil Biology and Fertility Institute of the International Centre
for Tropical Agriculture (TSBF-CIAT), P.O. 30677, Nairobi, Kenya
e-mail: b.vanlauwe@cgiar.org

J. Kihara et al. (eds.), *Improving Soil Fertility Recommendations in Africa using
the Decision Support System for Agrotechnology Transfer (DSSAT)*,
DOI 10.1007/978-94-007-2960-5_8, © Springer Science+Business Media Dordrecht 2012

simulating phenology and yield of the dual purpose varieties under different environments. Field trials for seven varieties were conducted across three sites in two seasons and data on phenology and management, soil characteristics and weather was collected and used in the CROPGRO model. A stepwise procedure was used in the calibration of the model to derive the genetic coefficients. Two sets of data from Kakamega and Kitale were used in calibration process while 2006 data for Kakamega and Msabaha, were used for evaluation of the model. The derived genetic coefficients provided simulated values of various development and growth parameters that were in good agreement with their corresponding observed values for most parameters. Model evaluation with independent data sets gave similar results. The differences among the cultivars were also expressed through the differences in the derived genetic coefficients. CROPGRO was able to accurately predict growth, phenology and yield. The model predicted the first flowering dates to within 2–3 days of the observed values, the first pod dates within 3 days of the observed values and yields within 5–300 kg ha^{-1} of the observed yields. The genetic coefficients derived in CROPGRO model can, therefore, be used to predict soybean yield and phenology of the dual purpose soybean varieties across different agro-ecological zones.

Keywords DSSAT • Cropping System Model CROPGRO-Soybean • Calibration • Evaluation • Promiscuous soybean varieties

Introduction

Soybean [*Glycine max (L). Merr.*] has been cultivated in Kenya by smallholder farmers since 1904 (Musangi 1997). However, varieties grown are low in soil improving characteristics and yield. Current initiatives to promote soybean cultivation in Kenya involve the screening of germplasm for adaptability in different agro-ecological zones and farming systems (Piper et al. 1998). A number of varieties developed at the International Institute for Tropical Agriculture (IITA) and screened in limited areas of western Kenya have shown promise in their ability to produce a large amount of biomass when sufficient P is available in the soil (Chemining'wa et al. 2004; Wanjekeche 2004). These have been referred to as promiscuous dual-purpose varieties, since they exhibit (1) low specificity for the effective Rhizobium strains, resulting in effective and efficient biological N fixation and (2) a more balanced partitioning of their aerial dry matter between grain and leaf biomass, leaving a net amount of leaf-N in the soil from which subsequent crops can benefit. The populations of *Bradyrhizobium japonicum* required for effective nodulation of soybeans are not endemic to African soils (Hadley and Hymnowitz 1973; Maingi et al. 2006). To avoid the need to inoculate soybean with *Bradyrhizobium japonicum,* soybean cultivars known as Tropical Glycine Cross (TGx) have been developed which nodulate with *Bradyrhizobium spp.* populations indigenous to African soils (Kueneman et al. 1984; Pulver et al. 1985; Abaidoo et al. 2000). Inclusion of these varieties in existing cropping systems in Kenya provides a potential to reduce the need for fertilizer N, through a more integrated management of soil fertility. However, soybean production and biomass partitioning between grain and leaf biomass varies between cultivars and fluctuate in response to agro-environmental conditions, soil characteristics, and management variables that modify the environment for the crop such as date and density of planting.

Soybean is sensitive to photoperiod and temperature through its entire life cycle. Yield potential of different varieties is therefore expected to change across altitudinal and rainfall gradients. Little is known about the performance of these dual purpose varieties across different soils and agro-ecological environments in Kenya and consistency across sites and seasons. Whereas field screening trials necessary to evaluate the performance of dual-purpose soybean varieties are costly and time-consuming, simulation modeling can help in identifying sets of most promising or best-bet varieties across agro-ecological conditions, to fine-tune for specific environments and farming systems, or to explore the potential of cultivars in new areas before establishing field trials.

Crop models can only be used successfully if the newly introduced varieties are well described (both genetically and phenologically) and their respective genetic coefficients made available. The performance of a soybean simulation model of soybean depends primarily on how well it predicts biomass, pod and seed growth. It is particularly important that the model simulates well the partitioning between vegetative and reproductive organs in dual-purpose varieties. Several soybean models have been developed (Meyer et al. 1979; Sinclair 1986). The CROPGRO is one of the crop simulation models that is included in the Decision Support System for Agro-technology Transfer (DSSAT) (Tsuji et al. 1994; Hoogenboom et al. 1999; Jones et al. 2003) and has been used in many applications around the world (Tsuji et al. 1994; Boote et al. 1998b). The model is physiologically based and simulates the productivity of soybean cultivars under various management and environmental conditions (Singh et al. 1994; Boote et al. 1998b; Kaur and Hundal 1999). Development of crops is associated with the attainment of growth stages. It is of two types: phase development such as anthesis, first pod occurrence, pod filling, physiological maturity and, morphological development such as biomass, Leaf Area Index, grain weight among others (Kumar et al. 2008). These traits make the model an attractive tool for crop improvement (Banterng et al. 2004).

Growth of a crop cannot be simulated without prior simulation of development of the crop. Different varieties are genetically different, that is, they attain growth stages at different times even when grown at the same location. The genetics (genetic characters) of the varieties therefore allows the crop to respond differently under different environments (air temperature, soil conditions, photoperiod among others). Therefore, genetic coefficients allow the model to predict differences in development, growth and yield among different cultivars when planted in the same environment as well as the differences in the behavior of growth and development of a single cultivar (Boote et al. 1998a). Genetic coefficients are therefore cultivar specific parameters used by the crop model to predict soybean daily growth and development responses to weather, soil characteristics and management actions.

There has been increased interest in modeling soybean in order to predict vegetative and reproductive development of different cultivars under various crop management and environmental conditions (Colson et al. 1995). However, predicting yield potential is difficult because of the wide ranges in yield, growth habit, and reproductive development of soybean cultivars (Cooper 1977; Blanchet et al. 1989). The objectives of this study were to estimate genetic coefficients for the CSM-CROPGRO – Soybean model from typical information provided by crop performance tests and to evaluate the performance of the model under Kenyan conditions.

Materials and Methods

Site Description

The trials were set at the Kenya Agricultural Research Institute (KARI) sites in Kakamega, Kitale and Msabaha. The KARI Kakamega site is situated in Kakamega District in western Kenya. The district covers an area of 1,395 km². The average population density is 495 persons per km². The district lies within altitude 1,250–2,000 m above sea level (m.a.s.l) with the average annual rainfall ranging from 1,250 to 1,750 mm p.a. The average temperature in the district is 22.5°C for most of the year. The site is on Latitude 0.28°N and Longitude 34°E. There are two main cropping seasons in the district characterized by long rains (March to June) and short rains (August to October). The annual rainfall ranges between 1,250 and 2,000 mm p.a.

The Kitale site is situated in KEPHIS Regional Office. Kitale (35°00′E; 1°01′N) is in Trans Nzoia District in the Rift Valley Province and falls within the Upper midland-4 (UM4) agro-ecological zone (Jaetzold et al. 2006). The Kitale area is a productive agricultural zone and produces mainly maize. At an altitude of 2,100 masl, annual precipitation is 1,000–1,300 mm with a dry season between December and February. Average minimum and maximum temperature is 9°C and 27°C, respectively (Jaetzold and Schmidt 1983; Jaetzold et al. 2006).

The Msabaha site is situated in Malindi in the Kenyan north coast. It is on Longitude 40.05°E and Latitude 3.27°S and at 91 m.a.s.l. The site experiences very high temperatures ranging from 28 to 32°C throughout the year and receives about 900–1,200 mm p.a. Msabaha and Kitale sites have extremes in terms of temperatures (altitude), soils and rainfall.

Experimental Design

Seven dual purpose soybean introductions (SB 3, SB 8, SB 9, SB 15, SB 17, SB 19, SB 20) and a local check (SB 23) (Table 8.1) were planted at each site on plots of 6 m × 2.35 m, with 0.45 × 0.1 m inter-row and interplant spacing for year 1 (2006) and year 2 (2007) respectively. The treatments were replicated three times at each site. Each plot was treated with 124 kg DAP ha⁻¹ fertilizer at planting.

Table 8.1 Variety codes as used by IITA and TSBF and their classification in terms of maturity period

TSBF code	IITA code	Maturity
SB 3	TGx 1835-10E	Early
SB 8	TGx 1895-33 F	Early
SB 9	TGx 1895-49 F	Early
SB 15	TGx 1889-12 F	Medium/Late
SB 17	TGx 1893-10 F	Medium/Late
SB 19	TGx 1740-2 F	Early
SB 20	TGx 1448-2E	Medium/Late
SB 23	–	Early

Source: Modified from IITA (2000); The SB is an adopted TSBF coding for the TGx (Tropical Glycine Cross)-IITA coding

Note that SB23 refers to Nyala, a locally used variety a commonly used local variety in kenya

Data Collection

Yield and phenological data were collected from each replicate and means taken per site. Weather data was recorded at a weather station located about 100 m from the trial plots in KARI station and less than 1 km from the KEPHIS trial plots for Kakamega and Kitale respectively. The data included daily rainfall, daily minimum and maximum temperatures and daily solar radiation. Solar radiation data was estimated from sunshine hours using Weatherman provided in the DSSAT v 4.2 model (Jones et al. 2003) in situations where solar radiation was not measured.

Establishment and Management of the Trials

The experimental fields were prepared by the removal of stubble and were subsequently hand ploughed. After marking the individual plots the field was leveled with the help of a hand drawn plank. The plots were marked and rows drawn on them. 124 kg DAP ha^{-1} fertilizer was applied in the plots. Soybean seeds were subsequently drilled manually with the hand in the line (rows) at a depth of 2–3 cm. Days to 50% flowering, 50% pod formation and 50% pod filling were recorded by observations in all replications. Days to maturity were recorded. The crops were weeded as required and pests controlled using pesticides. No top dressing was applied during the growing season and no supplemental irrigation was provided, using only natural rainfall.

Soil Data

Pits of 1 m by 1 m and 2 m deep were dug and soil sampled at a depth of 15 cm per layer interval through to 150 cm from each side of the wall for the sites. For each level the soil sampled from the four walls of the pit was bulked and mixed thoroughly and a sample taken for analysis. The samples were analyzed for total soil N where nitrate nitrogen was extracted using 2 M KCl and determined by Cadmium reduction (Dorich and Nelson 1984), while ammonium was determined in 2 M KCl extract by the salicylate-hypochlorite colorimetric method (Anderson and Ingram 1993). The sum of nitrate and ammonium gave the total inorganic nitrogen. A soil file compatible with the DSSAT input requirements was prepared using the SBuild program that estimates Drained Upper Limit (DUL), Lower Limit of Plant Extractable soil water (LL), Saturated Water content (SAT), Bulk Density (BD), Saturated Hydraulic conduct (SSKS) and Root Growth Factor (RGF) as a function of the clay, silt, sand and carbon content of different layers of the soil (IBSNAT 1989) (Table 8.2).

Calibration and Evaluation

Model calibration was first conducted for the soil parameters and then for genetic coefficients. For the calibration of soil parameters, soil sample data was used to calculate soil parameters for the entire profile and for each soil layer with the soil data retrieval program of DSSAT (Tsuji et al. 1994). Parameters that were obtained for each soil layer were saturated water content, drained upper limit, and the lower limit of plant-extractable water. The soil fertility factor for each profile was adjusted in the soil files after testing the model with the determined genetic coefficients.

The CSM-CROPGRO-Soybean model uses 15 genetic coefficients to define development and growth characteristics of a soybean cultivar (Table 8.3). To determine the genetic coefficients of the soybean varieties, the minimum data set collected was used as inputs in the standard format of DSSAT Version 4.2. Measured plant characteristics were used as initial coefficients. The calibration process is a systematic search of possible values that the model will use to be able to predict accurately the observed parameters. The procedures described by Mavromatis et al. 2001 were used. Candidate cultivars provided by the model were run in the sensitivity analysis mode shell from maturity group (MG) IV onwards (Boote 2008, personal communication) (Fig. 8.1). This is because materials in the tropical areas are bred to be less sensitive to photoperiod. The best varieties that predicted maturity close to the observed value were adopted for further adjustments.

Adjustments were then made for the EM-FL, FL-SH, FL-SD and the SD-PM in the sensitivity analysis shell in the model in order to get the best fit values which provided the least RMSE. These coefficients are measured in photothermal days. A photothermal day is the measure of the period to which temperatures are above the minimum required temperature for a particular crop to allow growth and development. This was done in an optimization shell which allowed different combinations either singly or in combination of the genetic coefficients to give results that best compared with the observed. The experimental data collected in the 2007 Kakamega KEPHIS experiment were used to create the crop management file (FileX) and the observed data (FileA), along with the weather and soil input files. The generic coefficients provided in the DSSAT model for various maturity groups (MG) were used as a starting point in the process of determining the genetic

Table 8.2 Characteristics of the different layers of the soil profile at the study sites

	Layer (cm)	LL (cm³ cm⁻³)	DUL (cm³ cm⁻³)	SAT (cm³ cm⁻³)	RGF (0–1)	SSKS (cm h⁻¹)	BD (g cm⁻³)	OC (%)	NI (%)	CL (%)	SI (%)	pH in H₂O
Kakamega	0–15	0.252	0.433	0.516	1.000	0.23	1.17	2.23	0.21	33	38.00	5.5
	15–30	0.241	0.404	0.493	0.638	0.23	1.24	1.82	0.21	33	34.00	5.4
	30–45	0.286	0.432	0.487	0.472	0.06	1.26	1.67	0.15	43	24.00	5.5
Kitale	0–15	0.214	0.353	0.462	1.000	0.43	1.33	1.57	29	26	0.15	5.4
	15–30	0.222	0.333	0.432	0.638	0.43	1.42	1.13	33	16	0.09	4.9
	30–45	0.228	0.337	0.426	0.472	0.12	1.44	0.99	35	16	0.07	5.1
Msabaha	0–15	0.145	0.223	0.366	1.000	0.43	1.62	0.43	0.04	23	2.00	6.8
	15–30	0.153	0.232	0.369	0.638	0.43	1.61	0.32	0.04	25	2.00	7.2
	30–45	0.169	0.253	0.377	0.472	0.43	1.59	0.20	0.03	29	4.00	7.2

The following characteristics were measured: *OC* percentage organic carbon, *CL* percentage clay, *SI* percentage silt and *NI* percentage nitrogen. These other characteristics were calculated by the model: *DUL* drained upper limit, *LL* lower limit of plant extractable soil water, *SAT* saturated soil water contact, *RGF* root growth factor, *SSKS* saturated hydraulic conductivity and *BD* bulk density

Table 8.3 Definition of genetic coefficients using in the CSM-CROPGRO-Soybean model

Parameter	Definition	Units
EM-FL	Time between plant emergence and flower appearance (R1)	PTD
FL-SH	Time between first flower and first pod (R3)	PTD
FL-SD	Time between first flower and first seed (R5)	PTD
SD-PM	Time between first seed (R5) and physiological maturity (R7)	PTD
FL-LF	Time between first flower (R1) and end of leaf expansion	PTD
SLAVR	Specific leaf area of cultivar under standard growth conditions	$cm^2 g^{-1}$
SIZLF	Maximum size of full leaf (three leaflets)	cm^2
WTPSD	Maximum weight per seed	g
For the following we used the default DSSAT values		
CSDL	Critical Short Day Length below which reproductive development progresses with no day-length effect (for short-day plants)	h
PPSEN	Slope of the relative response of development to photoperiod with time (positive for short-day plants)	1/h
LFMAX	Maximum leaf photosynthesis rate at 30 C, 350 vpm CO_2, and high light	$mg\ CO_2 m^{-2}\ s^{-1}$
XFRT	Maximum fraction of daily growth that is partitioned to seed + shell	
SFDUR	Seed filling duration for pod cohort at standard growth conditions	PTD
SDPDV	Average seed per pod under standard growing conditions	#/pod
PODUR	Time required for cultivar to reach final pod load under optimal conditions	PTD

PTD Photo-thermal days (Mavromatis et al. 2001)

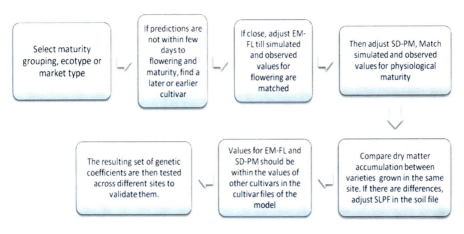

Fig. 8.1 Systematic approach that was used for calibration of the CSM-CROPGRO-Soybean model

coefficients (Grimm et al., 1993; Boote et al. 1997). The various default MGs (IV onwards) were tried until a default MG cultivar that predicted nearly the right total crop life cycle to maturity within 2–3 days was found (Table 8.4). Later or earlier MGs were also evaluated in case the predictions were not within a few days for anthesis and maturity. If close, EM-FL was adjusted until simulated and observed

Table 8.4 Modified genetic coefficients for the new varieties as determined using Kakamega 2007 data sets

| Variety name | ECO# | CSDL | PPSEN | EM-FL | FL-SH | FL-SD | SD-PM | FL-LF | LFMAX | SLAVR | SIZLF | XFRT | WTPSD | SFDUR | SDPDV | PODUR |
		1	2	3	4	5	6	7	8	9	10	11	12	13	14	15
IB0059 SB 15	SB0505	11.88	0.340	28.9	9	9.0	32.0	18	1.00	300	150	1	0.18	22.0	2.05	10.0
IB0058 SB 17	SB0504	12.83	0.303	27.0	10	14.0	33.0	18	1.00	300	150	1	0.18	23.0	2.05	10.0
IB0057 SB 19	SB0503	13.09	0.294	25.0	8	14.0	33.2	26	1.03	300	180	1	0.19	23.0	2.20	10.0
IB0056 SB 20	SB0502	11.88	0.340	28.9	7	13.5	31.5	15	1.03	300	180	1	0.18	22.0	2.05	10.0
IB0012 SB 23	SB0501	13.09	0.294	23.0	8	15.0	26.0	26	1.03	300	180	1	0.19	23.0	2.20	10.0
IB0062 SB 3	SB0508	13.09	0.294	25.0	9	13.8	30.0	25	1.00	300	190	1	0.19	25.5	2.40	10.5
IB0061 SB 8	SB0507	13.09	0.294	28.9	10	13.8	30.0	28	1.00	300	190	1	0.19	25.5	2.40	10.5
IB0060 SB 9	SB0506	12.58	0.311	28.9	9	11.0	25.0	18	1.00	300	150	1	0.18	22.0	2.05	10.0

Refer to Table 8.3 for definitions

values for flowering matched or were within a reasonable range. SD-PM was then adjusted to match observed physiological maturity (Fig. 8.1).

Adjustments for EM-LF and SD-PM were made such that the values were within the range of the other cultivars provided in CSM-CROPGRO-Soybean. In situations where the values were smaller or larger, it required adjustment for sensitivity to photoperiod by adjusting CSDL or PPSEN. Since there was no basis for adjusting these coefficients, it was necessary to go through the previous steps again.

FL-SD and SD-PM were adjusted together within the range supported by literature (Piper et al. 1996a). After fitting both FL-SD and SD-PM, FL-SH was set proportionally to FL-SD as follows:

$$FL - SH = \left(FL - SH / FL - SD\right)_{MG} FL - SD_{opt}$$

Where, $FL-SD_{opt}$ is the optimized value for FL-SD and the ratio of $(FLSH/FLSD)_{MG}$ is the ratio of the general maturity group values for the specific coefficients. This is to rescale the FL-SH so that the overall maturity period is maintained.

The default (generic) LFMAX and XFRT as provided by the MGs selected were maintained since they were not measured. This procedure was repeated for all the eight varieties. The SLPF in the soil file was set at 0.7, with 1.0 representing a very fertile soil, moving on down to predict the observed yield. Despite the output from the soils creation program, the SLPF for tropical soils was generally not set to be higher than 0.92 (the default value for Gainesville, FL) (Boote 2008, personal communication).

The accuracy of the procedure used to estimate the genetic coefficients was determined by comparing the simulated values of development and growth characters with their corresponding observed values, and by the values of root mean square error (RMSE) (Hunt et al. 1993; Tsuji et al. 1994). The derived genetic coefficients of individual varieties were compared to determine the sensitivity of the model and to capture the differences among these eight tested varieties in the experiment. The KEPHIS Kakamega 2007 and KEPHIS Kitale 2007 data sets were used in the calibration process. The sets of coefficients adopted from this procedure were then used for evaluation using the KEPHIS Kakamega 2006 and KEPHIS Msabaha 2006 data sets. This was intended to show how well the model predicted yield and phenology for other contrasting sites.

Results

Soil Profile Calibration

Soils for the three sites were sampled and analyzed for texture, organic carbon, total nitrogen and bulk density. The other parameters were generated by the soil data retrieval program of DSSAT. The results are shown in Table 8.2. The soil fertility factor was set at 0.7 reflecting the inherent soil fertility of these tropical soils.

Estimated Cultivar Coefficients

The varieties were selected from maturity group IV onwards from the cultivar files after running the model in the sensitivity analysis shell. The maturity groupings that gave 2–3 days difference to the observed maturity days were selected as potential cluster for the these varieties. SB 19, SB 23, SB 3, and SB 8 originated from MG IV, SB 15 and SB 20 from MG IX, SB 17- MG V, and SB 9 -MG VI. However, not all the varieties gave the desired 2–3 days difference between the simulated and observed maturity days (SB 17, SB 19, SB 3 and SB 8). This could partly be explained by the fact that maturity does not only depend on the maturity groupings (CSDL) but also other coefficients such as SD-PM. This was solved by optimizing the other coefficients and also bearing in mind the other parameters such as days to flowering, days to podding and yield.

The coefficients EM-FL, FL-SD, FL-SH and SD-PM were optimized until the RMSE between the simulated and the observed was lowest in the sensitivity analysis mode as described by Mavromatis et al. 2001 and Hoogenboom et al. 1999. The resulting genetic coefficients are shown in Table 8.4. The values for SLAVR measured from the field were very low; hence, the lowest value of 300 provided by the model was used for all the varieties. Other values for the rest of the coefficients were left as provided in the respective (generic values) maturity groupings (Table 8.4).

In assessing the accuracy of the genetic coefficients derived from model calibration, simulated values for four of the most critical developmental stages of the eight tested soybean varieties for the two growing seasons were compared with the corresponding observed values. Close agreements between observed and simulated values were obtained for days to first flowering and days to first pod production (Fig. 8.2). The model predicted the first flowering dates within 2–3 days of the observed values, and predicted the first pod dates within 3 days of the observed values (Table 8.5) for the calibration site (KEPHIS Kakamega 2007).

Evaluation of the CSM-CROPGRO-Soybean Model

The CSM-CROPGRO-Soybean model in DSSAT v 4.2 was evaluated using the data collected at KEPHIS Kakamega. After model calibration, based on the data from the Kakamega KEPHIS 2007 and Kitale KEPHIS 2007 experiments, the calibrated model was run using the calibrated genetic coefficients. The model was able to predict flowering, first pod and maturity dates for Kitale site within acceptable ranges (Table 8.6). The Kitale site, a high altitude area had RMSE for yield as low as 5 kg ha^{-1} and 187 kg ha^{-1} being the highest. Flowering and first pod was predicted very well with differences of up to 2 days and up to 3 days respectively (Table 8.6). The number of days to physiological maturity was also well predicted with up to a maximum of 5 days difference. Due to the high altitude the growing period for the varieties were highest compared to the other sites and this effect was well captured by the model. The model was then tested using a separate 2006 data set for the

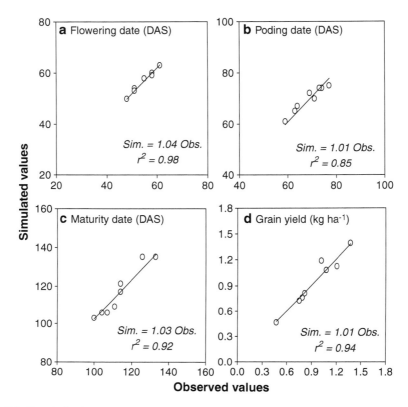

Fig. 8.2 Correlation between the simulated and the observed for various variables for all varieties for 2007 Kakamega data

Kakamega site. The results are tabulated in Table 8.7. The model predicted yield well for most of the varieties with RMSEs ranging from as low as 40 to less than 300 Kg/ha (Fig. 8.3). A correlation was done to determine the accuracy of prediction of the model (Fig. 8.4). The model was able to predict the flowering and the podding dates well. However, the yields were not well predicted.

Using KEPHIS Msabaha 2006 data set (a low altitude site) the model predicted the beginning of flowering and first pod date and yield very well (Table 8.8). Physiological maturity dates were not taken for the year but clearly the high temperatures substantially reduced the growing and maturity period for the varieties. There were large yield differences observed for SB 23, SB 3 and SB 8 (the early varieties).

The phenological stages were well simulated by the model for most of the varieties. The first flowering and first pod dates were simulated within an acceptable range of 2–5 days except for SB 3 which gave an error of 7 days. The canopy heights were overestimated by the model as compared by the measured values. Data for measured days to maturity were missing.

Table 8.5 Simulated vs. observed data for Kakamega 2007

Variety	Simulated flowering (DAS)	Observed flowering (DAS)	Simulated podding (DAS)	Observed podding (DAS)	Simulated maturity (DAS)	Observed maturity (DAS)	Simulated grain yield (kg ha⁻¹)	Observed grain yield (kg ha⁻¹)	RMSE (kg ha⁻¹)	Simulated canopy height (m)	Observed canopy height (m)	RMSE Canopy height (m)
SB 15	61	63	77	75	126	135	473	461	8	0.61	0.89	0.204
SB 17	55	58	69	72	114	121	788	757	31	0.55	0.53	0.107
SB 19	51	54	63	65	111	109	1,075	1,080	108	0.51	0.56	0.041
SB 20	61	63	74	74	133	135	1,374	1,395	82	0.62	0.56	0.127
SB 23	48	50	59	61	100	103	816	811	34	0.49	0.29	0.249
SB 3	51	53	64	67	107	106	1,017	1,189	309	0.51	0.3	0.236
SB 8	58	60	73	74	114	117	1,207	1,119	48	0.57	0.61	0.057
SB 9	58	59	71	70	104	106	751	724	28	0.56	0.94	0.286
LSD	0.82			0.84		2.52		129.7				–

*F test: variety = p < 0.001 at 5% level of significance

DAS days after sowing. RMSE gives the accuracy of model prediction

Table 8.6 Simulated vs. observed data for the varieties (of soybean) phenology and yield for Kitale 2007

Variety	Simulated flowering (DAS)	Mean observed flowering (DAS)	Simulated podding (DAS)	Mean observed podding (DAS)	Simulated maturity (DAS)	Mean observed maturity (DAS)	Simulated grain yield (kg ha^{-1})	Mean observed grain yield (kg ha^{-1})	RMSE (kg ha^{-1})
SB 15	68	70	86	87	136	140	1,107	1,065	5
SB 17	60	61	75	77	122	126	2,078	1,914	164
SB 19	56	57	68	70	117	121	2,244	2,123	109
SB 20	68	68	82	85	143	147	2,617	2,531	86
SB 23	51	53	64	64	105	108	1,613	1,667	73
SB 3	56	58	69	70	112	117	2,016	2,068	59
SB 8	64	64	79	82	122	126	2,551	2,623	187
SB 9	64	64	77	80	111	115	1,811	1,929	143
LSD	–	0.95	–	3.55	–	3.55	–	312.1	–

*F test: variety = p < 0.001 at 5% level of significance

RMSE root mean square error, *DAS* days after sowing. RMSE gives the accuracy of model prediction

Table 8.7 Simulated vs. observed data for Kakamega 2006

Variety	Simulated flowering (DAS)	Observed flowering (DAS)	Simulated podding (DAS)	Observed podding (DAS)	Simulated maturity (DAS)	Observed maturity (DAS)	Simulated grain yield (kg ha^{-1})	Observed grain yield (kg ha^{-1})	RMSE (kg ha^{-1})	Simulated canopy height (m)	Observed canopy height (m)	RMSE Canopy height (m)
SB 15	58	61	74	75	122	127	932	988	104	0.62	0.64	0.032
SB 17	52	54	66	68	111	120	1,091	1,194	103	0.54	0.41	0.108
SB 19	49	48	60	60	107	110	1,253	1,097	130	0.51	0.48	0.06
SB 20	58	60	71	69	129	133	1,682	1,407	244	0.62	0.47	0.155
SB 23	45	46	57	58	97	109	1,012	972	40	0.48	0.25	0.256
SB 3	49	55	62	65	103	112	1,187	1,417	230	0.51	0.39	0.202
SB 8	56	55	70	72	110	117	1,420	1,123	291	0.57	0.52	0.119
SB 9	56	56	68	69	100	109	973	873	96	0.56	0.72	0.117
LSD	–	0.85	–	0.98	–	4.5	–	310.5	–	–	0.26	–

*F test: variety = p<0.001 at 5% level of significance

RMSE root mean square error, *DAS* days after sowing levels. RMSE gives the accuracy of model prediction for the soybean yields

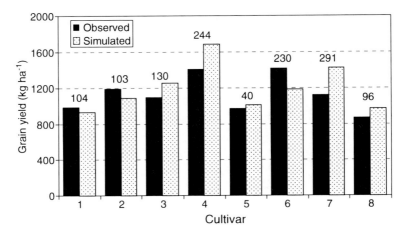

1- SB 15, 2-SB 17, 3- SB 19, 4- SB 20, 5- SB 23, 6- SB 3, 7- SB 8, 8- SB 9

Fig. 8.3 Comparison between simulated and observed yield for Kakamega site 2006

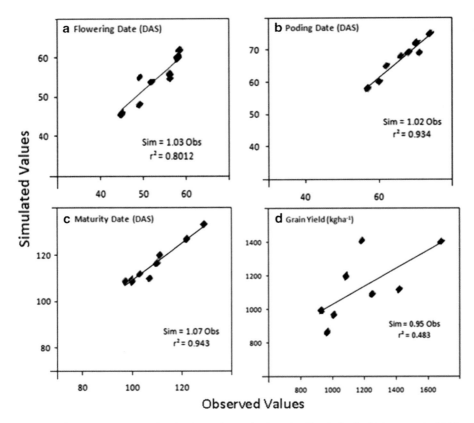

Fig. 8.4 Correlation between simulated and observed values for all varieties for Kakamega site 2006

Table 8.8 Simulated vs. observed data for the soybean varieties tested (phenology, yield and canopy height) for Msabaha 2006

Variety	Simulated flowering (DAS)	Mean observed flowering (DAS)	Simulated podding (DAS)	Mean observed podding (DAS)	Simulated maturity (DAS)	Observed maturity (DAS)	Simulated grain yield (kg ha⁻¹)	Mean observed grain yield (kg ha⁻¹)	RMSE (kg ha⁻¹)	Simulated canopy height (m)	Observed canopy height (m)
SB 15	36	38	47	45	86	–	468	552	96	0.6	0.42
SB 17	34	39	44	49	83	–	512	605	79	0.54	0.31
SB 19	32	38	40	44	81	–	656	716	55	0.49	0.37
SB 20	36	40	44	50	91	–	699	802	110	0.59	0.29
SB23 (Nyala)	29	31	38	35	72	–	515	231	216	0.48	0.23
SB 3	32	39	41	48	77	–	595	265	235	0.49	0.47
SB 8	36	39	46	47	81	–	664	315	256	0.56	0.43
SB 9	36	38	45	45	73	–	476	438	22	0.57	0.45
LSD	–	0.95	–	3.55	–	–	–	191.1	–	–	–

*F test: variety = p<0.001 at 95% significant levels. RMSE gives the accuracy of model prediction for the soybean yields

Table 8.9 Differences among soybean cultivars for the most important coefficients as determined by the calibration process

Variety	CSDL	EM-FL	FL-SH	FL-SD	SD-PM	FL-LF
SB 15	11.88	28.9	9	9.0	32.0	18
SB 17	12.83	27.0	10	14.0	33.0	18
SB 19	13.09	25.0	8	14.0	33.2	26
SB 20	11.88	28.9	7	13.5	31.5	15
SB 23 (Nyala)	13.09	23.0	8	15.0	26.0	26
SB 3	13.09	25.0	9	13.8	30.0	25
SB 8	13.09	28.9	10	13.8	30.0	28
SB 9	12.58	28.9	9	11.0	25.0	18
Mean	12.69	26.95	8.75	13.01	30.09	18.75
Standard Deviation	0.53	2.34	1.04	1.98	3.08	1.035
Coefficient of Variation	4.19	8.69	11.83	15.24	10.24	11.83

CSDL Critical Short Day Length below which reproductive development progresses with no daylength effect, *EM-FL* Time between plant emergence and flower appearance, *FL-SH* Time between first flower and first pod, *FL-SD* Time between first flower and first seed, *SD-PM* Time between first seed and physiological maturity, *FL-LF* Time between first flower and end of leaf expansion

Differences Among Cultivars

Differences among varieties for phenological coefficients were rather small, with the CV among varieties for individual characters ranging from 4.19% to 15.24% (Table 8.9). The greatest variation among the varieties was FL-SD followed by SD-PM. This is due to the high differences in time to first seed and maturity. It should be noted that despite the low CVs obtained, the varieties are different based on the physical appearances of the plants and the seeds. There were significant differences between SB 23, a local variety and the rest of the varieties. EM-FL and FL-SD were particularly distinct. SB 23 had the lowest and the highest values for EM-FL and FL-SD respectively compared to the rest of the varieties.

Discussion

The results clearly demonstrate that the CSM-CROPGRO-Soybean simulates phenology and yields quite well for most of the varieties. However, most critical is to obtain accurate genetic coefficients (Addiscot et al. 1995; Aggarwal et al. 1995) for the new varieties and soil parameters for the study sites (Heiniger et al. 1997). This involves collection of high quality data from the field. This means that the trials must be well managed with no water or nutrient limitations (Hodges and French 1985; Muchow 1985; Banterng et al. 2004) and also free from diseases. Otherwise, the coefficients resulting from such trials will not be of quality hence the subsequent simulations will result to large margins of error compared to the observed data.

The biggest challenge in the determination of genetic coefficients is the fact that the new varieties being developed or already developed must be well known. It is therefore critical for the modeler to get as much information on the varieties as possible. This is because some of the coefficients, especially those that determine the sensitivity to photoperiod, e.g. CSDL and PPSEN, are not easily determined. Hence, prior knowledge of the varieties is imperative.

The differences observed between the phenological stages and yield in Kakamega for the 2 years is attributed to the differences in temperature (Tables 8.5 and 8.7) (Butterfield and Morison 1992; Hoogenboom et al. 2004), rainfall, solar radiation and soil dynamics (Alagarswamy et al. 2000). The yield differences in Kakamega site in 2006 for observed and simulated values can also be attributed by small errors in the genetic coefficients especially for those attributes that were not measured. The status of the soil and the differences in the planting dates (Hoogenboom et al. 1999) also contributed to the differences in yield and phenology. The reliability of the genetic coefficients can only be guaranteed if they are generated from data derived from several years and several sites. Yield stability analysis could be performed over a wide range of environments and the nature of genotype × environment interactions could also be investigated. Ultimately this would improve the efficiency and effectiveness of variety evaluation and selection.

The ratio of sand to silt and clay affects the yield and phenology. This is due to the fact that these affect the water holding capacity of the soil and consequently affecting the growth and development of the plant. The model is designed to incorporate these effects while simulating growth.

The Kakamega, Kitale and Msabaha sites are all in the extremes in terms of temperature regimes. Kitale is the coldest; Kakamega is cold while Msabaha is very hot. This affects the flowering, pod filling and maturity durations of the soybean varieties across the sites (Piper et al. 1996b; Grimm et al. 1993) although the general trend as impacted by the maturity groups remains the same. The maturity period is significantly reduced at Msabaha compared to Kakamega and Kitale. High temperatures and high humidity in the coastal area tend to accelerate the photosynthesis process and this therefore causes fast growth hence reducing the growing period of the varieties. The growing period is longest in Kitale as a result of the low temperatures in the region.

The large yield differences between the observed and simulated yields in SB 23, SB 3 and SB 8 (the early varieties) at Msabaha were attributed to the fact that the varieties were harvested very late after they had long matured hence most of the grains were lost through shattering. However, the indeterminate nature of soybean also contributed to this because some pods were mature with grains while others were still very young in one plant, causing a delay in harvesting. Furthermore, the model does not consider the effects of pests and diseases (Banterng et al. 2004) and hence contributing to the overestimation of yields for this site.

SB 23, a local variety, also known as Nyala has unique genetic coefficients which are quite different from the other varieties. The CSDL is relatively the same as the early varieties, however the EM-FL, signifying the period between emergence and flowering is lowest compared to the other varieties. This is due to the fact that the dual purpose varieties tend to be more vegetative compared to the local checks. Therefore, substantial amount of time is taken to grow vegetatively for the dual purpose varieties unlike the SB 23. FL-SD is highest for SB 23 compared to the other varieties. FL-SD is the period between the first flower and first seed. The local varieties were primarily bred for yield as opposed to the dual purpose varieties. SB 23, therefore, takes more time in this stage of growth so as to maximize yields but may not necessarily have the highest yields depending on the other factors. The genetic coefficients clearly define the observed parameters for the different varieties. These parameters react with the environment to give different values of yield and phenological values for different agro-environments.

Conclusions

The model accurately predicted first flowering, first pod and yield to within 2–3 days, 3 days and 5–300 kg ha^{-1} respectively of the field observed values. Therefore the model can reliably be used for further simulations to gauge the performance of the introduced soybean varieties under different environments in Kenya. It is also possible to use this model for breeding new improved varieties of soybeans. However, there is need to include more characteristics from more sites to determine the soybean genetic coefficients in order to make it more robust and reliable.

Acknowledgements This research was funded by the TSBF soybean project in collaboration with KEPHIS. We would like to acknowledge the work done by KEPHIS and TSBF staff in ensuring the success of the field trials and this research as a whole. Special thanks to AfNET for sponsoring the DSSAT training and the provision of the software for use.

References

Addiscot T, Smith J, Bradbury N (1995) Critical evaluation of models and their parameters. J Envir Qual 24:803–807
Aggarwal PK, Kalra N (1994) Analyzing the limitation set by climatic factors, genotype, and water and nitrogen availability on productivity of wheat II. Climatically potential yields and management strategies. Field Crop Res 36:161–166
Aggarwal PK, Matthews RB, Kropff MJ (1995) Opportunities for the application of systems approaches in plant breeding. In: Aggarwal PK, Matthews RB, Kropff MJ, vanLaar HH (eds) SARP research proceedings. Institute International Rice Research, Los Banos, pp 135–144
Alagarswamy G, Singh P, Hoogenboom G, Wani SP, Pathak P, Virmani SM (2000) Evaluation and application of the CROPGRO-soybean simulation model in a vertic inceptisol. Agr Syst 63:19–32
Anderson J, Ingram J (1993) Tropical soil biology and fertility. A hand book of methods, 2nd edn. CABI, Wallingford, 221 pp
Banterng P, Patanothai A, Pannangpetch K, Jogloy S, Hoogenboom G (2004) Determination of genetic coefficients for peanut lines for breeding applications. Eur J Agron 21:297–310

Blanchet R, Bouniols A, Gelfi N, Wallace SU (1989) Response of determinate and indeterminate soybeans to fertilizing irrigation in soils of different depths. In: Pascale AJ (ed) Proceedings of 4th world soybean research conference, 5–9 March 1989, Buenos-Aires. Orientancion Grafica Editora Buenos Aires SRL, Buenos Aires, pp 740–745

Boote KJ, Tollenaar M (1994) Modeling genetic yield potential. In: Boote KJ, Bennett JM, Sinclair TR, Paulsen GM (eds) Physiology and determination of crop yield. ASA – CSSA – SSSA, Madison, pp 533–565

Boote KJ, Jones JW, Hoogenboom G, Wilkerson GG (1997) Evaluation of the CROPGRO-soybean model over a wide range of experiments. In: Kropff MJ et al (eds) Systems approaches for sustainable agricultural development: applications of systems approaches at the field level. Kluwer Academic, Boston, pp 113–133

Boote KJ, Jones JW, Hoogenboom G, Pickering NE (1998a) The CROPGRO model for grain legumes. In: Tsuji GY et al (eds) Understanding options for agricultural production. Kluwer Academic, Dordrecht, pp 99–128

Boote KJ, Jones JW, Hoogenboom G (1998b) Simulation of crop growth: CROPGRO MODEL. In: Peart RM, Curry RB (eds) Agricultural systems modeling and simulation. Marcel Dekker, New York, pp 651–692

Butterfield RE, Morison JI (1992) Modelling the impact of climate warming on winter cereal development. Agr Forest Meteorol 62:241–261

Cheminin'gwa GN, Muthomi JW, Obudho EO (2004) Screening of promiscuous soybean varieties in the North Rift valley of Kenya. Legume research network project newsletter. Issue no. 11

Colson J, Bouniols A, Jones JW (1995) Soybean reproductive development: adapting a model for European cultivars. Agron J 87:1129–1139

Cooper RL (1977) Response of soybean cultivars to narrow rows and planting rates under weed-free conditions. Agron J 69:89–92

Dorich RA, Nelson DW (1984) Evaluation of manual cadmium reduction methods for determination of nitrate in potassium chloride extracts of soils. Soil Sci Soc Am J 48:72–75

Grimm SS, Jones JW, Boote KJ, Hesketh JD (1993) Parameter estimation for predicting flowering dates of soybean cultivars. Crop Sci 33:137–144

Hadley HH, Hymnowitz T (1973) Speciation and cytogenetics. In: Caldwel BE (ed) Soyabeans: improvement production and uses. American Society of Agronomy, Madison, pp 97–116

Heiniger RW, Vanderlip RL, Welch WS (1997) Developing guidelines for replanting grain sorghum: I. Validation and sensitivity analysis of the SORKAM sorghum growth model. Agr J 89:5–83

Hodges T, French V (1985) Soybean growth stages modeled from temperature, daylength and water availability. Agron J 77:500–505

Hoogenboom G, Wilkens PW, Tsuji GY (eds) (1999) DSSAT version 3, vol 4. University of Hawaii, Honolulu

Hoogenboom G, Jones JW, Wilkens PW, Porter CH, Batchelor WD, Hunt LA, Boote KJ, Singh U, Uryasev O, Bowen WT, Gijsman AJ, du Toit A, White JW, Tsuji GY (2004) Decision support system for agro-technology transfer version 4.0 [CD-ROM]. University of Hawaii, Honolulu

Hunt LA, Pararajasingham S, Jones JW, Hoogenboom G, Imamura DT, Ogoshi RM (1993) GENCALC: software to facilitate the use of crop models for analyzing field experiments. Agron J 85:1090–1094

IBSNAT (International Benchmark Sites Network for Agrotechnology Transfer Project) (1989) Technical report 1.Experimental design and data collection procedure for IBSNAT. The minimum data sets for systems analysis and crop simulation, 3rd edn. University of Hawaii, Honolulu

IITA (2000) International Institute of Tropical Agriculture

Jaetzold R, Schmidt H (1983) Farm management handbook of Kenya: natural conditions and farm management information, vol II/A. Nyanza and Western Provinces, Ministry of Agriculture, Kenya in cooperation with German Agricultural Team (GAT) of GTZ, Kenya

Jaetzold R, Schmidt H, Hornetz B, Shisanya C (2006) Farm management handbook of Kenya: natural conditions and farm management information, vol II/C. Nyanza and Western Provinces, Ministry of Agriculture, Kenya

Jones JW, Hoogenboom G, Porter CH, Boote KJ, Batchelor WD, Hunt LA, Wilkens PW, Singh U, Gijsman AJ, Ritchie JT (2003) DSSAT cropping system model. Eur J Agron 18:235–265

Kaur P, Hundal SS (1999) Forecasting growth and yield of groundnut (Arachis hypogaea) with a dynamic simulation model PNUTGRO under Punjab conditions. J Agric Sci 133:167–173

Kueneman EA, Root WR, Dashiell KE, Hohenberg J (1984) Breeding soybean for the tropics capable of nodulating effectively with indigenous Rhizobium spp. Plant Soil 82:387–396

Kumar A, Pandey V, Shekh AM, Dixit SK, Kumar M (2008) Evaluation of CROPGRO-soybean (glycine max. [L] Merrill) model under varying environment condition. Ama Eurasian J Agron 1(2):34–40

Maingi JM, Gitonga NM, Shisanya CA, Hornetz B, Muluvi GM (2006) Population levels of indigenous bradyrhizobia nodulating promiscuous soybean in two Kenyan soils of the semi-arid and semi-humid agroecological zones. J Agr Rural Dev Trop Subtrop 107(2):149–159

Mavromatis T, Boote KJ, Jones JW, Irmak A, Shinde D, Hoogenboom G (2001) Developing genetic coefficients for crop simulation models with data from crop performance trials. Crop Sci 41:40–51

Meyer GE, Curry RB, Streeter JG, Mederski HJ (1979) SOYMOD/OARDC – a dynamic simulator of soybean growth, development and seed yield: I. Theory, structure and validation. Research Bulletin 1113. Ohio Agriculture Research and Development Center, Wooster

Muchow RC (1985) Phenology: seed yield and water use of grain legumes grown under different soil water regimes in a semiarid tropical environment. Field Crop Res 11:81–97

Musangi RS (1997) Soyabean cultivation in Kenya: status and prospects. Villa Maria Consultants, FAO Briefing paper

Piper EL, Boote KJ, Jones JW, Grimm SS (1996a) Comparison of two phenology models for predicting flowering and maturity date of soybean. Crop Sci 36:1606–1614

Piper EL, Smit MA, Boote KJ, Jones JW (1996b) The role of daily minimum temperature in modulating the development rate to flowering in soybean. Field Crop Res 47:211–220

Piper EL, Boote KJ, Jones JW (1998) Evaluation and improvements of crop models using regional cultivar trial data. Trans ASAE 14:435–446

Pulver EL, Kueneman EA, Ranga-Rao V (1985) Identification of promiscuous nodulating soybean efficient in nitrogen fixation. Crop Sci 25:660–663

Robert AC, Keyser HH, Singleton PW, Borthakur D (2000) Bradyrhizobia spp (TGx) isolates nodulating the new soybean cultivars in Africa are diverse and distinct from bradyrhizobia that nodulate north American soybeans. Int J Syst Evol Microbiol 50:225–234

Sinclair TR (1986) Water and nitrogen limitations in soybean grain production: I. model development. Field Crop Res 15:125–141

Singh P, Boote KJ, Virmani SM (1994) Evaluation of the groundnut model PNUTGRO for crop response to plant population and row spacing. Field Crop Res 39:163–170

Tsuji GY, Uehara G, Balas S (eds) (1994) DSSAT version 3, vol 1–3. University of Hawaii, Honolulu

Wanjekeche (2004) Screening of promiscuous soybean varieties in the north rift valley of Kenya. Legume research network project newsletter. Issue no. 11

Chapter 9
Simulation of Potential Yields of New Rice Varieties in the Senegal River Valley

Michiel E. de Vries, Abdoulaye Sow, Vincent B. Bado, and Nomé Sakane

Abstract Irrigated rice in the Sahel has a high yield potential, due to favorable climatic conditions. Simulation models are excellent tools to predict the potential yield of rice varieties under known climatic conditions. This study aimed to (1) evaluate new rice genotypes for the Sahel, and (2) calibrate simulation models to predict potential yield of irrigated rice in the Sahel. Two new inbred lines (ITA344 and IR32307) and one *O. sativa* × *O. glaberrima* line (WAS 161-B-9-2) were tested against IR64, an international check, and Sahel 108, locally the most popular rice cultivar. Field experiments were executed at two sites along the Senegal river, Ndiaye and Fanaye, differing in temperature regime and soil type. All cultivars were sown and transplanted at two sowing dates in February and March 2006. Observed grain yields varied from 7 to 10 t ha^{-1} and from 6 to 12 t ha^{-1} at Ndiaye and Fanaye, respectively. The number of days until maturity ranged from 119 to 158, depending on cultivar, sowing date and site. Experimental data of one sowing date was used to calibrate both the DSSAT and ORYZA2000 models. According to ORYZA2000, the same cultivars needed 400°Cd more in Fanaye than in Ndiaye to complete their cycle. ORYZA2000 simulated phenology well, but yield was underestimated. After calibrating DSSAT, different sets of genetic coefficients gave similar results. Genetic coefficients that reflected the observed phenology well resulted in lower than

M.E. de Vries (✉) • N. Sakane
Sahel station, Africa Rice Center, BP 96, St. Louis, Senegal

Plant Production Systems, Wageningen University, P.O. Box 430, 6700 AK
Wageningen, the Netherlands
e-mail: michielerikdevries@gmail.com

A. Sow • V.B. Bado
Sahel station, Africa Rice Center, BP 96, St. Louis, Senegal

J. Kihara et al. (eds.), *Improving Soil Fertility Recommendations in Africa using the Decision Support System for Agrotechnology Transfer (DSSAT)*,
DOI 10.1007/978-94-007-2960-5_9, © Springer Science+Business Media Dordrecht 2012

observed yields. Crop growth simulation is a powerful tool to predict yields, but local calibration at the same sowing date is needed to obtain useful results.

Keywords *Oryza sativa* • Crop growth simulation models • Irrigated rice • Sahel

Introduction

Irrigated rice production supplies a large portion of the national diets of Sahelian countries, and rice demand has been growing at 5.6% per annum (WARDA 2006). Yield potential of irrigated rice has been estimated at 8–12 t ha^{-1}, depending on cultivar, sowing date and site (De Vries et al. 2011; Dingkuhn and Sow 1997). High incident radiation levels and high temperatures create a favorable environment for irrigated rice cultivation, although a number of climatic constraints that may depress yields have been identified. The Africa Rice Center has created a new generation of genotypes from interspecific crosses between *O. sativa* japonica and *O. glaberrima*; tolerant to biotic and abiotic stresses in the African region (Jones et al. 1997). Notably NERICA genotypes (*Oryza sativa × O. glaberimma*) are a promising new source of germplasm for lowland conditions (Heuer et al. 2003; Saito et al. 2010; Sie et al. 2007). However, potential yield of these new varieties has not yet been quantified.

Rice-double cropping is physically possible in most Sahelian irrigation schemes, but in practice, less than 10% of the area is used for rice twice a year (Vandersypen et al. 2006). Reduction of growing cycle length of varieties, while keeping the potential yield at the same level, has been pointed out as a means to increase double-cropping acreage. To be able to plant twice a year and to avoid critical periods of heat and cold stress, farmers have to be aware of optimum sowing dates (Dingkuhn et al. 1995b; Poussin et al. 2003; Segda et al. 2005). Along the Senegal River, sowing dates for some cultivars have been optimized, and disseminated to farmers (Kebbeh and Miezan 2003). For newly generated germplasm, this information is not yet known.

To be able to determine optimum planting dates for the Sahelian zone, and to quantify the influence of climate on newly developed varieties, decision support tools are necessary. In climate change studies, crop growth simulation models are commonly used (Matthews et al. 1997; Xiong et al. 2009). Such tools have been developed for rice: e.g. RIZDEV (Dingkuhn et al. 1995a) and ORYZA1 (Kropff et al. 1994). A new generation of models have been developed that integrate water and nutrient limitation, including DSSAT 4 (Jones et al. 2003) and ORYZA2000 (Bouman et al. 2001). These models need to be locally calibrated in order to become useful for further research. ORYZA2000 has been evaluated for N-limitation (Bouman and Laar 2006) and water limitation (Feng et al. 2007) and for photoperiod sensitive varieties (Boling et al. 2011). It has been used under a variety of conditions in India (Arora 2006), Indonesia (Boling et al. 2010) and various sites across Asia (Jing et al. 2008). In two reviews (Timsina and Humphreys 2003; 2006a) showed that CERES-rice, a component of the DSSAT system was calibrated and evaluated using experimental data from more than one site or from more than one season only by Pathak et al. (2004). It has been used to simulate the rice-wheat system in India (Sarkar and Kar 2006; Saseendran et al. 1998) and for regional yield forecasts in China (Xiong et al. 2008). Both models have extensively been used in Asia, Australia and the America's, but up to now, not in Africa.

Although CERES-rice has only been partially described in different publications, it is relatively widely used (Timsina and Humphreys 2006b). For ORYZA2000, detailed calculations are described in Bouman et al. (2001) and these are freely available to researchers. The model is mainly developed to support scientific research. The new generation of decision support tools incorporates management factors other than planting date. This study aims to make a first step in evaluating the potential production of new varieties in comparison with simulation results of these two models as a way to predict yield and analyze the yield gap. Furthermore, we provide recent data, which can be used to improve crop performance at farmers' level in irrigated rice in Sahelian region of Africa.

Materials and Methods

Field Experiment

A field experiment was conducted to obtain data needed to calibrate the simulation models. A two-factor split plot experiment was set-up as follows: five rice varieties were planted at two sites, with three replicates; site was used as block, variety as plot. The varieties used were: (1) IR64, (2) WAS 161-B-9-2, an irrigated NERICA, a cross between IR64 and TOG5681, an *Oryza glaberrima* variety, (3) ITA344, (4) IR32307-107-3-2-2 and (5) Sahel 108 (IR13240-108-2-2-3).

The experiments were conducted at Ndiaye (16°11′N, 16°15′W) and Fanaye (16°32′N, 15°11′W) in Senegal. Both are experimental research stations of the Africa Rice Center (AfricaRice). The sites were located in the delta (Ndiaye), 35 km inland, and middle (Fanaye), 150 km inland, of the Senegal river valley. For a detailed description of the physical and chemical properties of the soils see Bado et al. (2008) and De Vries et al. (2010).

The experiments were established using a seed-bed, from which 21 day old seedlings were transplanted into a pre-soaked and leveled plot of 3×5 m. Seed beds were sown on 15 and 27 February 2006 in Ndiaye and Fanaye, respectively. Hill spacing was 0.2×0.2 m with two seedlings per hill. The level of standing water was kept constant at 5–10 cm. Fertilizer was applied at a rate of 120 kg N ha^{-1} and 21 kg P ha^{-1}. Nitrogen was applied in the form of urea and diammonium phosphate in three splits: 40% at early tillering, 30% at panicle initiation and 30% at booting stage. Phosphorus was applied as diammonium phosphate at early tillering. The plots were kept weed and pest free to ascertain potential growing conditions.

Phenology of all plots was observed; flowering was determined as the day 50% of the plants attained anthesis, and maturity as the day 80% of the pants were mature.

Table 9.1 Genetic coefficients for rice used in DSSAT 4

Code	Description
P1	Time period (expressed as growing degree days [GDD] in °C above a base temperature of 9°C) from seedling emergence during which the rice plant is not responsive to changes in photoperiod. This period is also referred to as the basic vegetative phase of the plant
P20	Critical photoperiod or the longest day length (in hours) at which the development occurs at a maximum rate. At values higher than P20 developmental rate is slowed, hence there is delay due to longer day lengths
P2R	Extent to which phasic development leading to panicle initiation is delayed (expressed as GDD in °C) for each hour increase in photoperiod above P20
P5	Time period in GDD°C) from beginning of grain filling (3–4 days after flowering) to physiological maturity with a base temperature of 9°C
G1	Potential spikelet number coefficient as estimated from the number of spikelets per g of main culm dry weight (less lead blades and sheaths plus spikes) at anthesis. A typical value is 55
G2	Single grain weight (g) under ideal growing conditions, i.e. non-limiting light, water, nutrients, and absence of pests and diseases
G3	Tillering coefficient (scalar value) relative to IR64 cultivar under ideal conditions. A higher tillering cultivar would have coefficient greater than 1.0
G4	Temperature tolerance coefficient. Usually 1.0 for varieties grown in normal environments. G4 for japonica type rice growing in a warmer environment would be 1.0 or greater. Likewise, the G4 value for *indica* type rice in very cool environments or season would be less than 1.0

At harvest, grain yield was measured from a 2×3 m surface. Meteorological data were recorded on-site, using Onset Hobo© weather stations. Data for temperature, solar radiation, wind speed and air humidity were recorded at hourly intervals from which daily values were derived.

Models

The generic and dynamic simulation model CERES-Rice which is part of the DSSAT system was used. The Cropping System Model (CSM) released with DSSAT v4.0 represents a major departure from previously released DSSAT crop models (Jones et al. 2003). The computer source code for the model has been extensively restructured into a modular format in which components separate along scientific disciplinary lines and are structured to allow easy replacement or addition of modules (Jones et al. 2003). It contains a detailed description of crop growth under optimal, nitrogen-limited, and water-limited conditions. The model operates on a daily time-step and calculates biomass production, which is then partitioned to the leaves, stems, roots and grain, depending on the phenological stage of the plant. The model uses genetic coefficients for different cultivars as model inputs to describe crop phenology in response to temperature and photoperiod (Boote and Hunt 1998). An overview of the genetic coefficients used for rice is given in Table 9.1.

The genetic coefficients can be divided into two categories. Firstly photothermal ones: P1 and P5, governing thermal time needed to complete a growth stage; and P20 and P2R, defining photoperiodism, and secondly morphological ones, G1, G2 and G3 defining number of spikelets, grain weight and tillering, and G4 which is a temperature coefficient. DSSAT4 has been calibrated for the cultivar IR64. Hence we used data from the experiments to calibrate and validate the performance of DSSAT4.

ORYZA2000 is a dynamic simulator of rice growth. Earlier versions comprise ORYZA1 (Kropff et al. 1994), ORYZA_W (Wopereis et al. 1994) and ORYZA_N (Ten Berge et al. 1997). ORYZA2000 simulates potential, water-limited and nitrogen-limited yield of lowland rice. The program is written in the FST language (Van Kraalingen et al. 2003), and the source code is made public on the web (http://www.knowledgebank.irri.org/oryza2000/). ORYZA2000 was calibrated as described in Bouman et al. (2001). In short, it uses observed phenological and climatic data to generate crop stage specific growth rates. Once calibrated for one site, it can be used to simulate crop growth at that site using climatic data as input. ORYZA2000 was calibrated for all varieties involved in the experiments. Data from the first sowing date were used for calibration and data of the second sowing date for validation.

Results

Climate

Meteorological data obtained at both experimental sites show that the climate at both sites is typical Sahelian (Fig. 9.1). Maximum temperatures were high but different: 46°C in Fanaye and 42°C in Ndiaye, and minimum temperatures increased over the growing season. Both temperatures and solar radiation were higher at Fanaye; no precipitation was recorded at either site during the season. Low minimum temperatures at the on-set of the growing season slowed down initial crop development.

Observed Grain Yield and Growing Cycle

Grain yields obtained at the two sowing dates at both sites are shown in Fig. 9.2 ranging between 5 and 11 t ha^{-1} (14% MC). Although temperatures and incident radiation were higher at the second sowing date, not all varieties performed better. IR64 and IR32307 performed better in Fanaye, notably IR64, with a yield of 11.1 t ha^{-1}, whereas ITA344 gave higher yields in Ndiaye, for the other varieties there were no differences between sites. In the Fanaye, the yields of all varieties were significantly higher at the second sowing date, except for IR64, while in

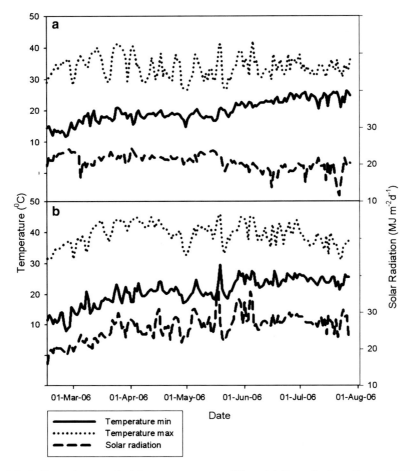

Fig. 9.1 Daily maximum and minimum temperatures (°C), and daily total solar radiation (MJ^{-1} m^{-2} day^{-1}) in Ndiaye (**a**) and Fanaye (**b**), Senegal in the dry season of 2006

Ndiaye yields at the second sowing date dropped to 4.5–5.8 t ha^{-1}. With the exception of Sahel 108, where no significant difference between sowing dates were observed. At the second sowing date in Ndiaye variation in the observations was higher. The yields obtained in Fanaye at the second sowing date can be regarded as potential yields, they varied between 9.8 and 11.1 t ha^{-1}. WAS161, IR64 and ITA344 recorded the highest yields. The performance of ITA344 in Ndiaye at the first sowing date can be explained by its cycle, it is the only medium duration variety in the experiment. It needed 158 days to complete its full cycle, whereas the other varieties needed between 132 and 141 days at the same site. In Fanaye, crop cycle was 119 and 121 days for IR32307 and IR 13240, respectively, and 126 days for WAS 161 and IR64, and 141 days for the medium duration ITA344. At Fanaye site the longer cycle of ITA344 did not result in increased yield. Growth rates over the complete growing season, calculated as grain yield over cycle, were between 58 and 70 kg

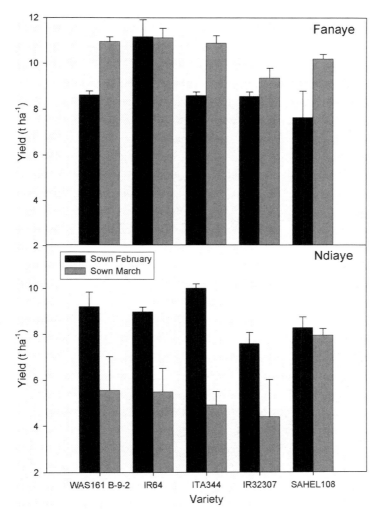

Fig. 9.2 Grain yield (t DM ha^{-1}) of five varieties at Ndiaye and Fanaye, Senegal, in 2006. *Black bars* represent a sowing date of 15 February and *grey bars* represent a sowing date of 14 and 16 March in Ndiaye and Fanaye, respectively. Error bars show standard error of mean

grain day^{-1}, at both sites and all varieties, although IR64 in Fanaye had a higher growth rate of 89 kg grain day^{-1}.

Simulation Results

The DSSAT model was parameterized with genetic coefficients of the rice variety IR64. The original coefficients as supplied with the software package were 500°Cd for P1, 450°Cd for P5 and 1.0 for G4 (Table 9.2). The original coefficients resulted

Table 9.2 Results of phenology (simulation of days to flowering and days to maturity), and grain yield (t DM ha^{-1}), using different sets of genetic coefficients (*P1*, *P5* and *G4*) for DSSAT4

Simulation set			Ndiaye			Fanaye		
			Flowering	Maturity	Yield	Flowering	Maturity	Yield
P1 (°Cd)	P5 (°Cd)	G4 (−)	(Das)	(Das)	(t ha^{-1})	(Das)	(Das)	(t ha^{-1})
200	450	1	46	81	5.7	44	74	2.7
500	450	0.75	68	103	8.7	57	85	6.3
500[a]	450	1	69	105	9.6	64	94	7.5
500	450	1.25	81	118	9.1	79	113	0
800	450	1	94	125	11.7	84	112	8.3
800	450	1.25	107	141	7.6	103	136	0
872	450	1	100	130	12.0	88	116	9.0
872[b]	600	1	100	137	12.4	88	123	10.2
872	700	1	94	137	12.2	84	123	8.6
872	450	1.25	113	147	5.3	108	142	0
950	450	1	106	136	12.4	94	121	8.5
1,000[c]	450	1	109	139	12.5	96	124	7.8
ORYZA2000	Calibrated		113	134	5.0	93	124	2.0
Observed			*100*	*136*	*9.4*	*95*	*126*	*11.1*

Results of ORYZA2000 after calibration, compared to observed results for variety IR64 in the dry season of 2006 in Ndiaye and Fanaye, Senegal

[a]DSSAT original genetic coefficients set for IR64

[b]Best performing set in Ndiaye

[c]Best performing set in Fanaye

in a short vegetative growing stage, 69 days simulated versus 100 observed, and a good yield estimation of 9.6 t ha^{-1} versus 9.4 t ha^{-1} observed, for Ndiaye. The same trend shows up in Fanaye, where the vegetative growth stage is simulated at 64 days, whereas 95 days were observed, and 7.5 t ha^{-1} was simulated versus 11.1 t ha^{-1} observed. Hence, the original crop coefficients were not successful in simulating the phenology, and, as all other processes are dependent on crop stage, the coefficients needed to be adapted (via calibration) in order to simulate the observed data. The coefficients that were chosen to be modified were P1 and P5 that govern the length of the vegetative and generative growth stage respectively, and G4, which regulates temperature responses (Table 9.1). Results of changes in P1, P5 and G4 values on time to flowering and maturity, and yield, using weather data from the Ndiaye and Fanaye sites, are presented in Table 9.2. Decreasing P1 decreased the duration from sowing to flowering, and similarly, decreasing P5 the duration from flowering until maturity. The two sets of genetic coefficients, which performed well (P1 = 872, P5 = 600 and G4 = 1 at Ndiaye and P1 = 1,000, P5 =450 and G4 =1 at Fanaye), were used to evaluate the simulation of biomass partitioning with DSSAT, see Fig. 9.3. The two sets of genetic coefficients were used at both sites. Simulation of stem weight shows a linear growth rate of stem weight up to 10 t ha^{-1} at Ndiaye and 12 t ha^{-1} at Fanaye. It shows that at the onset of grain filling about 50% of the stem weight is transformed into grain weight in one day.

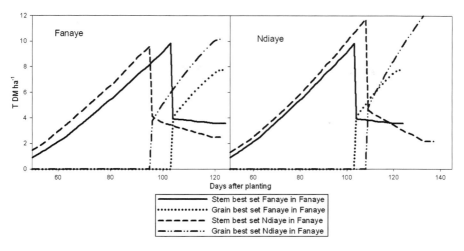

Fig. 9.3 Simulation of stem and grain weight by DSSAT 4 over time, using two calibrated sets of genetic coefficients; Best set Fanaye: P1 = 1,000 and P5 = 450, and best set Ndiaye: P1 = 872 and P5 = 600. Both sets were simulated using weather date of the 2006 dry season in Fanaye (*left*) and Ndiaye (*right*), Senegal. Simulation sets DSSAT

Grain filling rates are also different between the two sets and sites. Both models simulated biomass partitioning at a daily time step in a similar way. Biomass partitioning over time gives insight in whether models simulate processes realistically. As seen in the previous part, the models did not yet simulate grain yield satisfactory.

ORYZA2000, after being calibrated at each location using the DRATES program, simulated phenology well. However, simulated yields for cultivar IR64 were 5.0 and 2.0 t ha⁻¹, in Ndiaye and Fanaye, respectively, which was low compared to 9.4 and 11.1 t ha⁻¹ observed yields. To further test the performance of ORYZA2000, the model was calibrated for each variety at both locations, using data obtained at the first sowing date. The calibrated model was used to simulate the second sowing date, 1 month later. A calibrated ORYZA2000 simulated flowering in Ndiaye at 108 DAS, averaged over the five varieties, which is an underestimation of only 5 days, see Table 9.3. Maturity was simulated at 130 DAS, underestimating post-flowering stage by 5 days, to arrive at a total under-estimation of 10 days. In Fanaye, the pre-flowering period was on average overestimated by 10 days, whereas at maturity the difference was only 2 days. Yield simulation for Ndiaye resulted in yield values between 7.3 for IR64 and 9.7 t ha⁻¹ for ITA344. The observed yields were lower, by 2.7 t ha⁻¹ on average, but the model produced a good estimation of the potential yield. In Fanaye however, the model simulated low yields (6.3 t ha⁻¹ for ITA344 and 2.5 t ha⁻¹ for the other varieties), while the observed yields were very high: 9.4–11.1 t ha⁻¹.

Table 9.3 Validation of ORYZA2000: observed and simulated phenology (*DAS*) and yield (t ha⁻¹) of five rice varieties in Ndiaye and Fanaye (Senegal), sown at 14 and 16 March 2006, respectively

Site	Variety	Observed				Simulated			
		Phenology (DAS)			Yield (t DM ha⁻¹)	Phenology (DAS)			Yield (t DM ha⁻¹)
		PI	Flowering	Maturity		PI	Flowering	Maturity	
Ndiaye	WAS 161	57	111	141	5.6	57	108	127	7.4
	IR64	57	111	141	5.5	57	105	124	7.3
	ITA 344	59	129	151	4.9	59	123	149	9.7
	IR 32307	55	106	137	4.4	55	101	127	9.2
	Sahel 108	55	107	136	8.0	55	102	125	8.4
Fanaye	WAS 161	64	92	122	11.1	68	93	121	2.6
	IR64	59	88	122	10.9	69	99	123	2.4
	ITA 344	75	105	137	10.9	86	120	137	6.3
	IR 32307	57	84	120	9.4	61	94	117	2.6
	Sahel 108	61	85	120	10.2	64	96	119	2.5

For the simulation ORYZA2000 was calibrated for each location, using a data-set of an experiment that was sown 30 days before the simulated data-set

Parameter Sensitivity

The sensitivity of the DSSAT4 to changes of genetic coefficient was tested by changing the coefficients one by one (Table 9.2). By changing P1 from 500 to 200°Cd, the vegetative growth stage became even shorter, decreasing from 69 to 46 days. When coefficient G4 was changed, from 1.0 to 0.75, the phenology simulation for Ndiaye decreased by 2 days, but in Fanaye it decreased by 9 days, and yield decreased in both sites. When G4 was increased to 1.25, the cycle increased, from 105 to 118 days, but yield decreased by 481 kg ha^{-1} in Ndiaye. For Fanaye, the cycle increased by 9 days, but in all three cases, no yield was produced. The crop failure is probably due to high temperatures, for which sensitivity is defined by G4. To improve simulation results for Ndiaye, we increased factor P1, with the result that the vegetative stage increased to the desired length at 872°Cd. To simulate the generative phase, P5 was increased to 600°Cd. It gave good simulation results of phenology, but yield was overestimated by 2.9 t ha^{-1}. The same set of coefficients at the Fanaye site underestimated the vegetative growth stage by 7 days, and overestimated the generative phase by 4 days. The set of coefficients that accurately simulated phenology in Fanaye was 1,000 for P1, 450 for P5 and 1 for G4; the same set overestimated the complete cycle, and yield in Ndiaye.

For ORYZA2000, the genotypic parameters DVRI, DVRJ, DVRP and DVRR were calibrated using the DRATES program for variety Sahel108 in Fanaye, using the first sowing date. Then, the parameters were increased and decreased by 10%, and the difference in time to flowering and maturity using the original parameter set and the modified parameter set was calculated. The effect of an increase in the parameters was always – 2 days and a decrease resulted in +3 days for the number of days to flowering and maturity, except for DVRR, which had only an effect after flowering (Table 9.4). The standard deviations were small, and the effects were constant with sowing date. A change of 10% in a genotypic parameter results in a change of 2–3% in time to flowering or maturity. The combination of increasing or decreasing all parameters at the same time resulted in a change in time to flowering or maturity which was the sum of the individual effects. Hence, there were no interactions between genotypic parameters.

Discussion

The research aimed at finding potential yields of several new rice varieties, which were found to be 8–11 t ha^{-1}, which is in line with the performance of IR64 found by Dingkuhn and Sow (1997). The varieties performed differently at the two sites, due to climatic differences. For Ndiaye, ITA344, a medium duration, and WAS 161, a short duration NERICA can be recommended. For Fanaye, IR64 has the single best performance, while other tested varieties perform equally. Differences in temperature amplitude could have been the cause (Yin 1996). De Vries et al. (2011) show that the temperature amplitudes in Fanaye were larger than in Ndiaye.

Table 9.4 Sensitivity analysis for five genotypic parameters of ORYZA2000

Parameter	Δ	$\overline{\Delta}$ time to flowering (days)		Relative sensitivity[a]	$\overline{\Delta}$ time to maturity (days)		Relative sensitivity[a]
DVRI	+10%	−2	(0.5)	0.23	−2	(0.4)	0.18
	−10%	3	(0.6)	0.34	3	(0.7)	0.25
DVRJ	+10%	−2	(0.7)	0.27	−2	(0.6)	0.21
	−10%	3	(0.5)	0.38	3	(0.7)	0.30
DVRP	+10%	−2	(0.6)	0.21	−2	(0.5)	0.30
	−10%	3	(0.5)	0.30	3	(0.5)	0.23
DVRR	+10%	0	(−)	−	−2	(0.5)	0.22
	−10%	0	(−)	−	3	(0.5)	0.28
All combined	+10%	−6	(1.1)	0.78	−8	(1.4)	0.80
	−10%	8	(0.8)	0.98	11	(1.4)	1.00

DVRI development rate at initial growing stage, *DVRJ* development rate at juvenile growing stage, *DVRP* development rate at photoperiod sensitive growing stage, *DVRR* development rate at reproductive growing stage. All parameters have been changed 10% and the difference with the base runs of SAHEL 108 sown in Fanaye on 15 sowing dates has been calculated. The average difference in time to flowering and to maturity between the base runs and runs with modified parameters is shown. Between brackets is the standard deviation. Sensitivity of the parameters has been determined as the difference in input parameter over the difference in response variable.
[a]Relative sensitivity is calculated as (Δ parameter/default parameter value)/(Δ response variable/default variable value)

From Table 9.2 it can be concluded that the genetic coefficients that are supplied with the DSSAT4 program do not produce satisfactory results, underestimating the time to flowering by 30% at both sites. The genetic coefficients used as default values in DSSAT4 did not produce satisfactory results under Sahelian conditions. It underestimated the time to flowering by 30% at both sites. When the program was calibrated at the site, it did not simulate rice yield accurately, hence not all crucial growth processes were simulated correctly. In Ndiaye it overestimated rice yield by 32%, and in Fanaye it underestimated yield by 30%. When calibration sets of parameters for Fanaye and Ndiaye were used for the other site, the phenology simulation was not satisfactory. An explanation could be that the effects of extreme temperatures on yield are not simulated adequately. From the model description it is not clear how heat and cold stress are simulated, but in light of their importance in yield determination in the Sahel, they should be emphasized in future efforts to improve the model. Our results support the conclusions from Timsina and Humphreys (2006a), that DSSAT4 requires local calibration for each variety and each sowing date. Thus, DSSAT4 is not suitable for large scale explorations, as genotypic parameters, assumed to be constant, vary inter environment.

Simulation of rice under Sahelian conditions using ORYZA2000 resulted in an underestimation of yield (Tables 9.2 and 9.3). The high temperatures, well above 35°C (Fig. 9.1), could have resulted in simulation of a higher heat induced sterility. For ORYZA2000, calibration was done at each site and simulation results were validated using sowing date 30 days later. Phenology was underestimated by

ORYZA2000 (Table 9.3), while the temperature at the second sowing date was higher, indicating that under high temperatures development rate is slower than simulated by ORYZA2000. As Sheehy et al. (2006) have pointed out, temperature responses of the model are not always accurate. Yin et al. (1997) proposed an elegant method to simulate development rate using a non-linear approach, which could improve simulation results in environments with extreme temperatures such as the Sahel. When the model was validated, it overestimated yield in Ndiaye of variety Sahel 108 by between 5%, which can be used as a benchmark for potential yield of this variety for a March sowing in the delta area of the Senegal river. Bouman and Van Laar (2006), Belder et al. (2007) and Boling et al. (2007) showed that ORYZA2000 was well suited to simulate nitrogen and water limited rice growth under tropical Asian conditions, hence we can assume that the performance of ORYZA2000 under Sahelian conditions was largely affected by climate rather than by either nitrogen or water limitations. We can conclude that ORYZA2000 can be used to predict the phenology and yields of these new varieties but the model needs to be calibrated for each site and sowing date. There is need for further research to increase performance of simulation models as there are still essential processes in rice phenology apparently influenced by extreme temperatures ($T > 35°C$), which are not yet modeled satisfactorily. An increased understanding of the physiological basis of these processes will undoubtly result in increased model performance.

Acknowledgement   The authors are grateful to the Dutch Directorate General for International Cooperation (DGIS) of the Dutch government and Africa Rice Centre for funding this research.

References

Arora VK (2006) Application of a rice growth and water balance model in an irrigated semi-arid subtropical environment. Agric Water Manag 83:51–57

Bado BV, De Vries ME, Haefele S, Wopereis MCS, N'diaye MK (2008) Critical limit of extractable phosphorous in a gleysol for rice production in the Senegal river valley of West Africa. Commun Soil Sci Plant Anal 39:202–206

Belder P, Bouman BAM, Spiertz JHJ (2007) Exploring options for water savings in lowland rice using a modelling approach. Agric Sys 92:91–114

Boling AA, Bouman BAM, Tuong TP, Murty MVR, Jatmiko SY (2007) Modelling the effect of groundwater depth on yield-increasing interventions in rainfed lowland rice in Central Java, Indonesia. Agric Sys 92:115–139

Boling AA, Tuong TP, van Keulen H, Bouman BAM, Suganda H, Spiertz JHJ (2010) Yield gap of rainfed rice in farmers' fields in Central Java, Indonesia. Agric Sys 103:307–315

Boling AA, BoumanBAM, TuongTP, KonboonY, Harnpichitvitaya D (2011) Yield gap analysis and the effect of nitrogen and water on photoperiod-sensitive jasmine rice in north-east Thailand NJAS – Wageningen. J Life Sci 58:11–19

Boote KJ, Hunt LA (1998) Data for model operation, calibration, and evaluation. In: Tsuji GY, Hoogenboom G, Thornton PK (eds) Understanding options for agricultural production. Kluwer Academic, Dordrecht, pp 9–39

Bouman BAM, van Laar HH (2006) Description and evaluation of the rice growth model ORYZA2000 under nitrogen-limited conditions. Agric Sys 87:249–273

Bouman BAM, Kropff MJ, Tuong TP, Wopereis MCS, ten Berge HFM, van Laar HH (2001) Oryza2000: modeling lowland rice. International Rice Research Institute, Los Banos

De Vries ME, Rodenburg J, Bado BV, Sow A, Leffelaar PA, Giller KE (2010) Rice production with less irrigation water is possible in a Sahelian environment field. Crop Res 116:154–164

De Vries ME, Leffelaar PA, Sakané N, Bado BV, Giller KE (2011) Adaptability of irrigated rice to temperature change in Sahelian environments. Exp Agric 47:69–87

Dingkuhn M, Sow A (1997) Potential yields of irrigated rice in the Sahel. In: Miézan KM et al (eds) Irrigated rice in the Sahel: prospects for sustainable development. WARDA, Bouaké, Côte d'Ivoire, pp 361–379

Dingkuhn M, Le GalMPY, Poussin JC (1995a) RIDEV: un modèle de développement du riz pour le choix des variétés et des calendriers. In: Boivin P et al (eds) Nianga, laboratoire de l'agriculture irriguée en moyenne vallée du Sénégal ORSTOM, Atelier ORSTOM-ISRA, Saint-Louis, Senegal, pp 205–222

Dingkuhn M, Sow A, Samb A, Diack S, Asch F (1995b) Climatic determinants of irrigated rice performance in the Sahel – I. Photothermal and micro-climatic responses of flowering. Agric Sys 48:385–410

Feng L, Bouman BAM, Tuong TP, Cabangon RJ, Li Y, Lu G, Feng Y (2007) Exploring options to grow rice using less water in northern China using a modelling approach: I field experiments and model evaluation. Agric Water Manag 88:1–13

Heuer S, Miezan KM, Sie M, Gaye S (2003) Increasing biodiversity of irrigated rice in Africa by interspecific crossing of *Oryza glaberrima* (Steud.) x *O. sativa indica* (L.). Euphytica 132:31–40

Jing Q, Bouman BAM, van Keulen H, Hengsdijk H, Cao W, Dai T (2008) Disentangling the effect of environmental factors in yield and nitrogen uptake of irrigated rice in Asia. Agric Sys 98:177–188

Jones MP, Dingkuhn M, Aluko GK, Semon M (1997) Interspecific Oryza sativa L. x O. Glaberrima Steud. progenies in upland rice improvement. Euphytica 94:237–246

Jones JW, Hoogenboom G, Porter CH, Boote KJ, Batchelor WD, Hunt LA, Wilkens PW, Singh U, Gijsman AJ, Ritchie JT (2003) The DSSAT cropping system model. Eur J Agron 18:235–265

Kebbeh M, Miezan KM (2003) Ex-ante evaluation of integrated crop management options for irrigated rice production in the Senegal river valley. Field Crop Res 81:87–94

Kropff MJ, van Laar HH, Mattews RB (1994) ORYZA1: an ecophysiological model for irrigated rice production. DLO-Staring Centrum, Instituut voor Onderzoek van het Landelijk Gebied (SC-DLO), Wageningen

Matthews RB, Kropff MJ, Horie T, Bachelet D (1997) Simulating the impact of climate change on rice production in Asia and evaluating options for adaptation. Agric Sys 54:399–425

Pathak H, Timsina J, Humphreys E, Godwin DC, Bijay-Singh V, Shukla AK, Singh U, Matthews RB (2004) Simulation of rice crop performance and water and N dynamics, and methane emissions for rice in northwest India using CERES rice model. CSIRO Land and Water, Griffith

Poussin JC, Wopereis MCS, Debouzie D, Maeght JL (2003) Determinants of irrigated rice yield in the Senegal river valley. Eur J Agron 19:341–356

Saito K, Azoma K, Sie M (2010) Grain yield performance of selected lowland NERICA and modern Asian rice genotypes in West Africa. Crop Sci 50:281–291

Sarkar R, Kar S (2006) Evaluation of management strategies for sustainable rice-wheat cropping systems, using DSSAT seasonal analysis. J Agric Sci 144:421–434

Saseendran SA, Singh KK, Rathore LS, Rao GSLHVP, Mendiratta N, Narayan KL, Singh SV (1998) Evaluation of the CERES-rice version 3.0 model for the climate conditions of the state of Kerala, India. Meteorol Appl 5:385–392

Segda Z, Haefele SM, Wopereis MCS, Sedogo MP, Guinko S (2005) Combining field and simulation studies to improve fertilizer recommendations for irrigated rice in Burkina Faso. Agron J 97:1429–1437

Sheehy JE, Mitchell PL, Ferrer AB (2006) Decline in rice grain yields with temperature: models and correlations can give different estimates. Field Crop Res 98:151–156

Sie M, Kabore KB, Dakou D, Dembele Y, Segda Z, Bado BV, Ouedraogo M, Thio B, Ouedraogo I, Moukoumbi YD, Ba NM, Traore A, Sanou I, Ogunbayo SA, Toulou B (2007) Release of four new interspecific varieties for the rainfed lowland in Burkina Faso. Int Rice Res Notes 32:16–17

Ten Berge HFM, Thiyagarajan TM, Wopereis MCS, Drenth H, Jansen MJW (1997) Numerical optimization of nitrogen application to rice. Part I. Description of MANAGE-N. Field Crop Res 51:29–42

Timsina J, Humphreys E (2003) Performance and application of CERES and WAGMAN destiny models for rice-wheat cropping systems in Asia and Australia: a review 16/03. CSIRO land and water, Griffith, Australia

Timsina J, Humphreys E (2006a) Performance of CERES-rice and CERES-wheat models in rice-wheat systems: a review. Agric Sys 90:5–31

Timsina J, Humphreys E (2006b) Applications of CERES-rice and CERES-wheat in research, policy and climate change studies in Asia: a review. Int J Agric Res 1:202–225

Van Kraalingen DWG, Rappoldt C, van Laar HH (2003) The Fortran simulation translator, a simulation language. Eur J Agron 18:359–361

Vandersypen K, Keita ACT, Kaloga K, Coulibaly Y, Raes D, Jamin JY (2006) Sustainability of farmers' organization of water management in the office du Niger irrigation scheme in Mali. Irrig Drain 55:51–60

WARDA ARC (2006) Annual report 2005–2006: providing what's needed, Cotonou, Benin

Wopereis MCS, Bouman BAM, Kropff MJ, Berge HFMt, Maligaya AR (1994) Water use efficiency of flooded rice fields I. Validation of the soil-water balance model SAWAH. Agric Water Manag 26:277–289

Xiong W, Holman I, Conway D, Lin E, Li Y (2008) A crop model cross calibration for use in regional climate impacts studies. Ecol Model 213:365–380

Xiong W, Conway D, Lin E, Holman I (2009) Potential impacts of climate change and climate variability on China's rice yield and production. Clim Res 40:23–35

Yin X (1996) Rice flowering in response to diurnal temperature amplitude. Field Crop Res 48:1–9

Yin X, Kropff MJ, Horie T, Nakagawa H, Centeno HS, Zhu D, Goudriaan J (1997) A model for photothermal responses of flowering in rice. I. Model description and parameterization. Field Crop Res 51:189–200

Chapter 10
Modeling Maize Response to Mineral Fertilizer on Silty Clay Loam in the Northern Savanna Zone of Ghana Using DSSAT Model

Mathias Fosu, S.S. Buah, R.A.L Kanton, and W.A. Agyare

Abstract Maize has become the most important cereal in northern Ghana serving as cash crop and the main staple for most communities. It is the crop that receives most fertilizer input by farmers in the region, although the recommendations being used are over 20 years old. These recommendations were derived from relationships between crop yields and different applied rates of nutrient obtained from field fertilizer experiments. In most cases the experiments do not take care of the full extent of spatial and temporal variability associated with crop production. In addition the experiments are often very expensive and time consuming as they have to be carried out over many years in varied ecologies to be able to make valid recommendations for a region. Computer simulation model is a useful tool in this regard in reducing cost and time required for such studies, and also taking care of the spatial and temporal variability in the production of the crop. Response of maize to nitrogen in the northern Guinea savanna agro-ecology of Ghana was evaluated using the Seasonal Analysis component of the Decision Support System for Agrotechnoloy Transfer (DSSAT 4.02) – Cropping System Model (CSM). A simulation was performed for crop growth, development and yield of maize run for a site at Nyankpala near the University for Development Studies in the northern savanna agro-ecology. A field trial consisting of five nitrogen rates (0, 30, 60, 90, 120, kg/ha) with 30 kg K_2O and 30 kg P_2O_5/ha was simulated for 24 years using measured daily weather and soil records for the site. The field trial was conducted under rain-fed conditions on a silty

M. Fosu (✉)
Scientific Support Group, SARI, CSIR – Savanna Agricultural Research Institute,
P.O. Box 52, Tamale, Ghana
e-mail: m_fosu@hotmail.com

S.S. Buah • R.A.L. Kanton
Upper West Farming Systems Research Group, SARI, CSIR – Savanna Agricultural
Research Institute, P.O. Box 52, Tamale, Ghana

W.A. Agyare
Department of Agricultural Engineering, KNUST, Kumasi, Ghana

J. Kihara et al. (eds.), *Improving Soil Fertility Recommendations in Africa using the Decision Support System for Agrotechnology Transfer (DSSAT)*,
DOI 10.1007/978-94-007-2960-5_10, © Springer Science+Business Media Dordrecht 2012

clay loam soil (Gleyi-ferric luvisol) in 2006. Quality protein maize (QPM) variety (Obatanpa) was the test crop. The economic analysis of the model took account of weather- and price-related risks, after carrying out a strategic analysis. The results showed that increasing levels of N up to 120 kg/ha increased maize grain yield at a diminishing return. The model accurately simulated maize grain yield up to 90 kg/ha nitrogen application but failed to accurately predict maize grain yield when nitrogen was applied at 120 kg/ha. Excessive water stress induced by high N application negatively affected the growth of maize. Nitrogen was leached most at N application rate of 120 kg/ha. Maize production was not profitable at N application rate ≤ 30 kg/ha on silty clay loam of the Guinea Savannah zone of Ghana. DSSAT-CSM can be used to accurately predict maize growth, development and yield in Ghana if well calibrated.

Keywords Nutrient response • Maize • Northern Ghana • DSSAT model

Introduction

Maize is a major staple in sub-Saharan Africa and the most important cereal in Ghana (Fosu et al. 2004). It is grown in all the six agro-ecologies of the country. Even in the drier areas of northern Ghana where until a few years ago sorghum and millet were the dominant cereals, maize has gradually taken over both as cash crop and an important staple. About 793,000 ha are devoted to the crop country-wide with an average yield of 1.5 Mt/ha (SRID 2007). In southern Ghana where soils are relatively rich in nutrients, it is possible to obtain grain yields of 1.5–2 Mt/ha without soil amendment. In northern Ghana however, the soils are low in N (< 0.08% total N) and P (< 10 ppm Bray-1 P) giving very low maize yield without soil amendment, ranging between 0.2–0.8 Mt/ha (Ahiabor et al. 2007). The most limiting nutrient is N (Fugger 1999). Mineral fertilizer application is thus a major component of maize production in northern Ghana and N is frequently applied in greater amounts than any other nutrient.

Nitrogen level of 64 kg/ha has been recommended for maize production in northern Ghana but this blanket recommendation is over 20 years old (Obeng et al. 1990). Some attempts have been made in recent times to update the old recommendations. However, results of these attempts have been largely inconclusive as most of the trials could not run their full courses before the end of donor funding under which most of them were initiated. The common approach used in estimating the amount of fertilizer N needed by a crop involves the measurement of yield response to increasing rates of N in field experiments. For the results to be valid for use across different agro-ecologies and different soils, the trials need to be carried out over several years on major benchmark soils. This is time consuming and expensive. Also, the response of crops to N is rarely the same from year to year because this nutrient is dynamic and mobile. At best, equations from variable N rate experiments needed to determine the amounts of N fertilizer, describe only historical relationships in the data thus offering little insight into processes that must be understood to better manage N inputs (Bowen and Baethgen 1998).

In recent times crop models have been used to support field agronomic research in predicting the yield of many genotypes in different soils under different management regimes and different weather conditions provided requisite data are available (Boote et al. 1996). Crop models need to be tested and validated for a given site using long-term historical weather data to simulate crop performance under varying cultural practices such as sowing dates, sowing densities, cultivar selection, soil fertility and diseases (Naab et al. 2004). One of such models is the Decision Support System for Agro-technology Transfer (DSSAT).

DSSAT was developed by an international team of scientists to estimate production, resource use and risks associated with different crop production practices. Jones et al. (1998) described in detail the DSSAT Model. To use the model one needs to conduct field experiment on one or more crops and collect minimum data set required for running and evaluating a crop model. Soil and historical weather data for the site in the region need to be entered before performing simulation of new management practices to predict performance and uncertainty associated with each practice. Crop models are rarely used in sub-Saharan African countries due largely to lack of adequate knowledge about them, among many other reasons (Bontkes et al. 2001; Bontkes and Wopereis 2003).

The objectives of the experiment conducted at Nyankpala were (1) to test the performance of DSSAT in Guinea savanna agro-ecology of Ghana and (2) assess the current nitrogen recommendations for maize in northern Ghana.

Materials and Methods

Site Characteristics

The experiment was carried out in 2006 at University for Development Studies (UDS, N 09° 24. 767'; W 000° 59.358'; 183 m above sea level) near Tamale, Ghana. The site falls within the Guinea savanna zone of Ghana where mean annual temperature is about 28°C with day maximum reaching 42°C during the hottest months of February to March. The lowest temperature of about 20°C is recorded in the months of December and January. Relative humidity is 40–50% but can be as low as 9% during the afternoon in the driest months of November to March (Walker 1962). Rainfall is monomodal and the annual mean is about 1,100 mm with a variability of 15–20% (Kasei 1988). The soil was a shallow silty clay loam alfisol (about 60 cm deep), gravelly and concretionary with a plinthic horizon at 50 cm depth.

Field Experiment

The selected site was disc-ploughed and harrowed in June to a depth of 15 cm after composite soil samples were taken. The treatments were five levels of nitrogen (0, 30, 60, 90, 120 kg/ha) with each N level (except the zero) receiving 30 kg/ha of phosphorus (P_2O_5) and potassium (K_2O) in a randomized complete block design replicated three times. Quality protein maize (QPM) variety, Obatanpa (110 days to maturity) was sown by hand on June 17 to a depth of 5 cm at a spacing of 0.75 m by 0.40 m. Each plot was 15 m long and 7.5 m wide with 10 rows of maize separated by 1.0 m alleys and replicates were separated by 2.0 m alleys. The planted field was sprayed with pre-emergence herbicide atrazine 1 day after maize was sown at the rate of 4 L/ha using a hand operated knapsack sprayer. Subsequent weeding was by hand hoeing.

Two weeks after sowing (WAP), half of each N rate was applied as sulphate of ammonia together with the full rates of P as triple superphosphate and K as KCl banded and incorporated at about 4 cm depth. The other half of N was applied at 6 WAP and the field ridged with hand hoe.

Measurements

A soil profile pit was dug to a depth of 1.5 m and samples taken at 15 cm interval to a depth of 60 cm as uniform plinthite layer was observed from 60 to 150 cm. Soil water content at these depths were determined by gravimetric methods and converted to volumetric water content using bulk densities of each layer.

The sampled soils were analyzed at Savanna Agricultural Research Institute laboratory for particle size distribution (hydrometer method) ammonium and nitrate (colorimetric), total N (Kjeldahl), available P (Bray-1), exchangeable K (Flame photometry).

Number of days to 50% emergence, anthesis, silking and maturity were determined.

At maturity, the center six rows in each plot were harvested to determine grain and stover weight after air drying. Daily rainfall, maximum and minimum temperatures and sunshine hours were measured at a weather station about 300 m away from the site. Solar radiation (MJ/m^2) was estimated from sunshine hours using the WeatherMan utility programme of DSSAT v.4 (Wilkens 2004). Data on soil, climate, crop growth and yield were entered in the standard file formats (Soil.sol, *.WTH, *.MZX) needed to run the model.

Model Calibration

Model calibration was undertaken to ensure that the model worked well for the cultivar used (Hunt and Boote 1998).

Soil Water Balance and Soil Water Holding Characteristics

In the DSSAT model, volumetric water content in each soil layer varies between a lower limit (LL) to which plants can extract soil water and a saturated upper limit (SAT) as described by Ritchie (1985). If water content of a given layer is above the drained upper limit (DUL), i.e. water content at field capacity, then water is drained to the next layer with a "tipping bucket" concept using a drainage coefficient specified for the soil.

The water content at field capacity or DUL, lower limit of plant available soil water (LL), saturated upper limit (SAT), saturated hydraulic conductivity (K_{sat}) and root growth factor (RF) for each soil layer were initially estimated by entering soil particle size distribution, bulk density, organic matter and gravel contents into the soil file creation utility (SBuild).

Calibration of Genetic Coefficients

The DSSAT Model was calibrated for genetic coefficients of the maize variety Obatanpa (Hunt and Boote 1998). As these coefficients were not available for the maize variety used in the experiment, the medium season maize variety in the cultivar file MZCERO40.CUL in the DSSAT v4 model was used as starting point from which to calibrate Obatanpa. Genetic coefficients were determined by manipulating model simulation against data as described by Boote et al. (1998). Coefficients for duration to flowering (EM-FL), duration from flowering to seed (FL-SD) and duration from seed to maturity (SD-PM) were adjusted to predict the observed life cycle.

Simulation Experiment

The simulation experiment was performed with a model of crop growth, development and yield of maize module (Jones et al. 2003) run for the site at the University for Development Studies, Nyankpala near Tamale. A seasonal analysis was performed with the five N rates (0, 30, 60, 90, 120, kg/ha) replicated 24 times (years) using measured daily weather and soil data for the site, with quality protein maize (QPM) var. Obatampa. The nitrogen was split applied. Phosphorus and Potassium were basally applied at 30 kg P_2O_5/ha and 30 kg K_2O/ha. The field trial was conducted under rain-fed conditions on a silty clay loam to validate the model (Chude et al. 2001).

The seasonal analysis module was used to evaluate N losses through leaching and the maize response to the different N fertilizer rates. An economic analysis was also carried out to quantify net returns for each treatment, taking yield and product price risk into account. The price of maize grain was fixed at prevailing market price of GH¢0.20 per kg and the price of fertilizer was GH¢0.50 per kg.

After obtaining the distribution of economic returns, strategic analysis was carried out to compare treatments in economic terms, taking into account weather- and price-related risks. This is done by examining the mean-variance plots of gross margins or net returns per hectare, or using the mean-Gini stochastic dominance.

Mean-Variance (EV) analysis

For two risk prospects A and B, with means E(.) and variances V(.), respectively, then A dominates B if

$$E(A) = E(B) \text{ and } V(A) < V(B)$$

or if

$$V(A) = V(B) \text{ and } E(A) > E(B)$$

Mean-Gini Stochastic Dominance (MGSD)

For two risky prospects A and B, A dominates B by MGSD if

$$E(A) > E(B)$$

or if

$$E(A) - \int(A) > E(B) - \int(B)$$

where E(·) is the mean, and ʃ(·) the Gini coefficient of distributions A and B. ʃ is half the value of Gini's mean difference. It is a measure of the spread of a probability distribution.

The most economically superior treatment was then selected by this process.

Results and Discussion

Soil Water

The estimated soil water content at field capacity was below 0.2 cm^3 cm^{-3} for soil depth below 15 cm (Table 10.1). The lower limit of plant available soil water was approximately 0.1 cm^3 cm^{-3} for all soil layers. Saturated hydraulic conductivity was very slow for the 0–15 cm depth, moderate for 15–30 cm depth and slow and very slow for the 30–45 cm and 45–60 cm depths, respectively, according to classification by Landon (1991).

Weather

The weather data for 2006 are presented in Table 10.2. The mean annual temperature and solar radiation for 2006 were normal for the location but the rainfall was

Table 10.1 Soil characteristics: lower limit of plant available water (*LL*), drained upper limit (*DUL*), saturation (*SAT*), saturated hydraulic conductivity (*Ksat*), and root growth factor (*RF*) used for model simulation in Nyankpala, 2006

Soil depth	LL (cm³ cm⁻³)	DUL (cm³ cm⁻³)	SAT	K_{sat}	RF
0–15	0.115	0.176	0.324	0.43	1.000
15–30	0.099	0.148	0.280	2.59	0.638
30–45	0.071	0.132	0.186	1.32	0.472
45–60	0.063	0.099	0.168	0.43	0.350

Table 10.2 Weather data for Nyankpala, 2006

Month	Rainfall (mm)	Minimum temperature (°C)	Maximum temperature (°C)	Solar radiation (MJ m⁻²d⁻¹)
Jan.	0.0	21.7	37.5	19.7
Feb.	9.9	22.8	38.4	20.5
Mar.	8.7	25.9	37.7	18.3
Apr.	65.4	24.7	36.7	20.7
May	106.0	23.7	32.9	20.3
Jun.	105.9	23.4	31.7	19.7
Jul.	142.7	23.5	30.9	18.9
Aug.	107.5	22.9	29.9	17.3
Sep.	147.1	22.5	30.0	16.0
Oct.	131.2	23.0	31.5	20.2
Nov.	0.0	20.0	35.2	19.4
Dec.	0.0	17.9	35.3	16.6
	Total = 824.4	Mean = 22.7	Mean = 34.0	Mean = 19.0

only 75% of long-term mean. Dry spell (i.e., total consecutive daily rainfall < 4 mm) more than 7 days occurred in June, and more than or equal to 10 days occurred in July and August during the year. These conditions could result in water deficit with implications for reduced crop performance. Mean monthly minimum temperature was 26°C in March but remained below 24°C from May to December. The mean maximum monthly temperature for the growing period (June to October) was 30.8°C and the mean solar radiation for the same period was 18.4 MJ m⁻²d⁻¹.

Prediction of Grain Yield

The model accurately simulated maize grain yield up to 90 kg/ha nitrogen application but failed to accurately predict maize grain yield when nitrogen was applied at 120 kg/ha (Fig.10.1).

The failure of the model to predict accurately the grain yield of maize when 120 kg/ha of N was applied could be attributed to water stress at high N concentration. The soil at the experimental site is gravelly and fairly shallow with water content at field capacity below 0.2 cm³ cm⁻³.

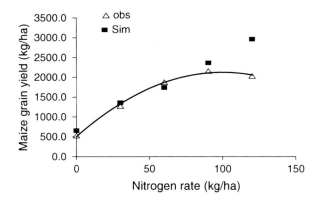

Fig. 10.1 Simulated and observed maize grain yield at Nyankpala as affected by nitrogen fertilizer applied in 2006. The root mean square error was 378.2

At all stages of the maize development, water stress was most severe when 120 kg/ha N was applied (Fig. 10.2). This is exacerbated by the gravelly nature of the soil resulting in a dry upper limit <0.25 cm^3 cm^{-3} of soil water in all soil layers (Table 10.1).

The model also showed that there was leaching of N at all N rates. There was a positive correlation between N applied and N leached with the highest N leached occurring at N application rate of 120 kg/ha (Fig. 10.3).

The increase in N leaching with increasing rates of application is expected in a soil having a sandy loam texture. This finding is similar to the observation by Powlson (1997) that application of large amounts of nitrogen results in leaching of nitrates and possible contamination of ground water (Addiscott 1996).

Economics of Fertilizer Application

The yield of maize observed and simulation for 2006 and 24 years for N rates of 0–120 kg/ha and monetary returns based on 2006 prices are presented in Table 10.3. The economic analysis of the long-term simulated maize grain yield indicated that maize production is not profitable at N rate of 30 kg/ha or lower. Based on the 2006 prices for maize grain and fertilizer, applying N at 120 kg/ha is the most profitable for the 24-year simulation (Table 10.3). Cumulative probability analysis for harvested grain yield indicated that 50% of the times, maize grain yield will be 500, 1,200, 2,000, 2,750 and 2,900 kg/ha for 0, 30, 60, 90 and 120 kg/ha N, respectively (Fig. 10.4).

The cumulative probability of monetary returns (Fig. 10.5) indicated that 75% of the time, for N rates of 0, 30, 60, 90, 120 kg/ha, the corresponding monetary returns will be $98, $0, $140, $250, and $300, respectively.

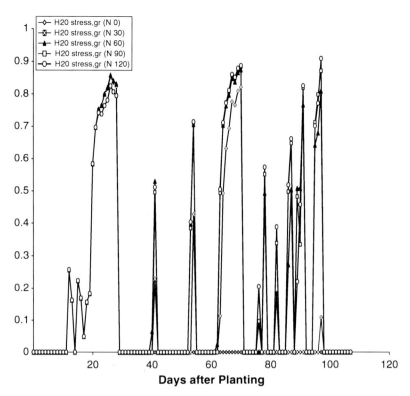

Fig. 10.2 Effect of N levels on water stress in soils at Nyankpala, 2006

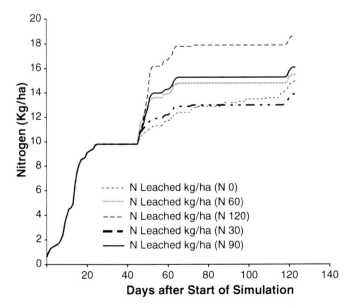

Fig. 10.3 Simulated nitrogen leached at different N rates during the growing season

Table 10.3 Observed and simulated yields of maize and monetary returns at various N rates at Nyankpala

Nitrogen rate (kg/ha)	Observed grain yd, 2006 (kg/ha)	Simulated grain yd, 2006 (kg/ha)	Simulated grain yd, 24 year (kg/ha)	Monetary returns- 24 year sim (US $):E(x)	E(x)-T(x)
0 N	529.4	653.0	448.2	−128.2	−143.9
30 N	1,273.6	1,360.3	1216.8	−36.0	−76.0
60 N	1,876.4	1,740.2	1994.5	74.6	8.5
90 N	2,158.3	2,366.8	2486.4	154.9	66.8
120 N	2,031.9	2,962.4	2824.0	179.6	79.4

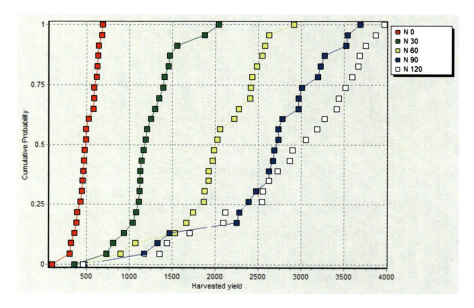

Fig. 10.4 Cumulative probability of maize yield per ha at different N rates

Conclusion

Seasonal Analysis using DSSAT Cropping System Model (CSM) of field experiment conducted to test the response of maize to five N levels in the Guinea savanna agro-ecology of Ghana showed that increasing levels of N increase maize grain yield but at a diminishing return. The model accurately simulated maize grain yield up to 90 kg/ha nitrogen application but failed to accurately predict maize grain yield when nitrogen was applied at 120 kg/ha. Excessive water stress induced by high N application negatively affected the growth of maize at this application rate. More N was leached at application rate of 120 kg/ha than all the other application rates.

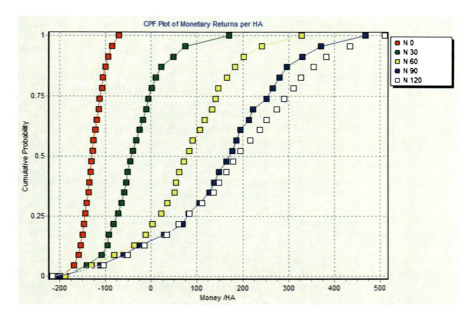

Fig. 10.5 Cumulative probability of monetary returns per hectare at different N rates

However, application of N at 120 kg/ha is profitable for maize production in the long term based on the simulated data in silty clay loams of the Guinea Savannah zone of Ghana. DSSAT-CSM can be used to accurately predict maize growth, development and yield in Ghana if well calibrated.

References

Addiscott TM (1996) Fertilizers and nitrate leaching. Issues Environ Sci 5:1–26
Ahiabor DK, Fosu M, Tibo I, Sumaila I (2007) Comparative nitrogen fixation, native arbuscular mycorrhiza formation and biomass production potentials of some grain legume species in the field in the Guinea savanna zone of Ghana. W Afr J Appl Ecol 11:89–98
Bontkes TES, Wopereis MCS (eds) (2003) Decision support tools for smallholder agriculture in Sub-Saharan Africa: a practical guide. IFDC-CTA publication, Muscle Soals
Bontkes TES, Singh U, Chude VO (2001) Problems and opportunities in adopting system tools for decision-making related to soil fertility improvement in Africa. Paper presented at the INMR symposium, CIAT, Cali, Columbia, 2001
Boote KJ, Jones JW, Pickering NB (1996) Potential uses and limitations of crop models. Agron J 88:704–716
Boote KJ, Jones JW, Hoogenboom G, Pickering NB (1998) The CROPGRO model for grain legumes. In: Tsuji GY, Hoogenboom G, Thornton PK (eds) Understanding options for agricultural production. Kluwer Academic, Dordrecht
Bowen WT, Baethgen WE (1998) Simulation as a tool for improving nitrogen management. In: Tsuji GY et al (eds) Understanding options for agricultural production. Kluver Academic, Boston, pp 189–204

Chude VO, Amapu IY, Yusuf AA, Singh U, Breman H, Dejean P (2001) Validation of CERES-maize model under sub-humid tropical conditions. Nig J Soil Res 2:37–42

Fosu M, Kühne RF, Vlek PLG (2004) Improving maize yield in the Guinea Savanna zone of Ghana with leguminous cover crops and PK fertilization. J Agron 3:115–121

Fugger W-D (1999) Evaluation of potential indicators for soil quality in savanna soils in northern Ghana. (West Africa). PhD Thesis, Georg-August Universität, Göttingen, Germany

Hunt LA, Boote KJ (1998) Data for model operation, calibration, and evaluation. In: Tsuji GY et al (eds) Understanding options for agricultural production. Kluver Academic, Boston, pp 9–39

Jones JW, Tsuji GY, Hoogenboom G, Hunt LA, Thornton PK, Wilkens PW, Imamura DT, Bowen WT, Singh U (1998) Decision support system for agrotechnology transfer: DSSAT v3. In: Tsuji GY et al (eds) Understanding options for agricultural production. Kluver Academic, Boston, pp 157–177

Jones JW, Hoogenboon G, Porter CH, Boote KJ, Batchelor WD, Hunt LA, Wilkens PW, Singh U, Grijsman AJ, Ritchie JT (2003) The DSSAT cropping system model. Eur J Agron 18:235–265

Kasei CN (1988) The physical environment of semi-arid Ghana. In: Unger PW, SneedTV, Jordan WR, Jensen R (eds) Challenges in dryland agriculture : a global perspective, Proceedings of the international conference on dryland farming, 15–19 Aug 1988, Amarillo/Bushland. Texas Agricultural Experiment Station, Texas, pp 350–354

Landon JR (1991) A booker tropical soil manual: handbook for soil survey and agricultural land evaluation in the tropics and subtropics. Booker Tate, Thame, pp 58–125

Naab JB, Piara Singh K, Boote J, Jones JW, Marfo KO (2004) Using the CROPGRO-peanut model to quantify yield gaps of peanut in the Guinean Savanna Zone of Ghana. Agron J 96:1231–1242

Obeng HB, Erbynn KB, Asante EO (1990) Fertilizer requirements and use in Ghana. Report submitted to the Government of Ghana. Tropical Agricultural Development Consultancy. Accra Ghana, pp 197

Powlson DS (1997) Integrating agricultural nutrient management with environmental objectives – current state and future prospects. Proceedings No. 402. International Fertilizer Society, York, pp 1–44

Ritchie JT (1985) A user-oriented model of the soil water balance in wheat. In: Fry E, Atkin TK (eds) Wheat growth and modeling. NATO-ASI Ser. Plenum Publ. Corp, New York, pp 293–305

SRID (2007) Facts and figures. Statistics, Research and Information Department of Ministry of food and Agriculture, Ghana

Walker HO (1962) Weather and Climate. In: Wills J (ed) Agriculture and land-use in Ghana. Oxford University Press, London

Wilkens PW (2004) DSSAT v4. Weather data editing program, WeatherMan. DSSAT v4 user's guide to cropping system models, vol 2. International Centre for soil fertility and agricultural development. University of Georgia, Griffin, Georgia, University of Florida, Gainsville

Chapter 11
Beyond Biophysical Recommendations: Towards a New Paradigm

Andre Bationo, Job Kihara, and Akin Adesina

Abstract African soils have an inherently poor fertility because they are very old and lack volcanic rejuvenation. Inappropriate land use, poor management and lack of input have led to a decline in productivity, soil erosion, salinization and loss of vegetation. The extent of such losses is of sufficient importance that action, such as recapitalization of soil fertility, increased use of inorganic fertilizer, and more efficient recycling of biomass within the farming system are being taken. For this reason, several long-term soil fertility management trials have been conducted in Africa to enable appropriate agronomic recommendations. Despite the need for increasing productivity, the average intensity of fertilizer use in SSA, excluding South Africa, is about 9 kg/ha. The diagnostic studies of fertilizer use in Africa have suggested that fertilizer use is low in Africa for four interrelated reasons: (1) The low returns to fertilizer use due to agro-climatic conditions and current farming methods; (2) The lack of information about fertilizer among retailers, farmers, and extension agents such as price information and best practices; (3) The high costs of fertilizers due to foreign production, large units, and costly transport; and (4) The inconsistent and adverse policy environment such as shifting government and donor subsidy policies that undermine private investment. This calls for a shift in paradigm to ensure adoption of the appropriate technologies emanating from

A. Bationo (✉)
Soil health program, Alliance for a Green Revolution in Africa (AGRA), Accra, Ghana
e-mail: abationo@agra-alliance.org

J. Kihara
AfNet Coordination, Tropical Soil Biology and Fertility (TSBF) Institute
of CIAT, c/o ICRAF, Nairobi, Kenya

A. Adesina
Alliance for a Green Revolution in Africa (AGRA), Nairobi, Kenya

long-term experiments and to realize the much needed green revolution in Africa. A critical lesson from previous paradigms is that a highly context-specific approach is required which takes into account the fertility status of the soil, the availability of organic inputs and the ability to access and pay for mineral fertilizers. However, making the necessary investments in soil fertilization to derive benefits including adequate returns on investments depends on output markets and the market value of farm products. This varies across Africa, within regions and even within villages and fields.

The Need for a Green Revolution

Inappropriate land use, poor management and lack of input have led to a decline in productivity, soil erosion, salinization and loss of vegetation. Most soils of Africa are poor compared to most other parts of the world. In addition to low inherent fertility, African soils nutrient balances are often negative indicating that farmers mine their soils. During the last 30 years, soil fertility depletion has been estimated at an average of 660 kg N ha^{-1}, 75 kg P ha^{-1} and 450 kg K ha^{-1} from about 200 million ha of cultivated land in 37 African countries. Through such soil nutrient depletion, Smaling et al. (1993) estimated that Africa lost \$4 billion per year. As a result of the inherent low fertility of African soils and subsequent land degradation, only 16% of the land has soil of high quality and about 13% has soil of medium quality. Fifty five percent of the land in Africa is unsuitable for any kind of cultivated agriculture except nomadic grazing. These are largely the deserts, which include salt flats, dunes and rock lands, and the steep to very steep lands, and about 30% of the African population (or about 250 million) are living or dependent on these land resources. Despite these farming conditions, many African economies are heavily dependent on agriculture with over 70% of the employment and a contribution of 40% to the national incomes.

Africa has been facing a serious food crisis characterized by over three decades of silent hunger, and malnutrition is projected to rise over the next 20 years. During the 2003–2005 period, food deficit of undernourished population in 45 African countries was 264 kcal/person/day compared to 174 for the rest of the world (FAO stat 2008). Food imports in Africa have risen by \$2.6 billion between 2006 and 2007 and are projected to rise from the current \$6.5 billion to \$11 billion by 2020 thereby draining the scarce resources in Africa. The economic, social and political costs of the food crisis are high and pose significant threats to the good economic growth that African countries have achieved over the past decade. The net impact of the high food prices is highest for poor net buyers of food, mainly the majority of African nations. In Africa, more than 300 million people live on less than a dollar per day. Sub-Saharan Africa has the highest proportion of population living in extreme poverty in the world, and this proportion has remained high for the last 25 years whereas significant poverty reductions occurred in other regions of the world over the same period, e.g., South and East Asia (World bank 2009). Many of the poor people are at risk of falling deeper into poverty as they spend between 50% and 60% of their incomes on food. The rapidly growing population, at 2.5% per annum (UN data 2006), strains the already overstretched food supply systems. The population of sub-Saharan Africa is projected to grow from 600 million in 2000 to nearly a billion by 2020 and International Food Policy Research Institute (IFPRI) projections indicate that Africa is the only region of the world that will continue to face major food shortages beyond 2020.

Soil-fertility depletion in smallholder farms is now widely accepted as the fundamental biophysical root cause of declining per capita food production in Africa (Sanchez et al. 1997) and is the main constraint to the achievement of Green Revolution. While the cumulative loss of crop productivity from land degradation worldwide between 1945 and 1990 has been estimated at 5%, as much as 6.2% of productivity has been lost in SSA. The degradation is mainly through water erosion (46%) and wind erosion (36%); other are loss of nutrients (9%), physical deterioration (4%) and salinization (3%) primarily driven by overgrazing (49%) and inappropriate agricultural practices (24%) as well as deforestation (14%) and overexploitation of vegetative cover (13%). Through desertification alone, for example, Africa is burdened with a US \$9.3 billion annual cost. Overall, an estimated US \$42 billion in income and 6 billion hectares of productive land are lost annually due to land degradation and resultant decline in agricultural productivity. African soil nutrient balances are largely negative indicating that farmers mine their soils and this is estimated to result in loss of about \$4 billion per year. Due to prevailing government policies, availability of information and limited access and affordability of fertilizers, the use of fertilizers in sub-Saharan Africa is the lowest in the world, estimated at only 8 kg ha^{-1} in year 2002, i.e., only 10% of the worlds' average (Maatman et al. 2008; Morris et al. 2007). As a result of the above problems, scientists have concluded that soil fertility replenishment should be considered as an investment in natural resource capital, and a key to solving the chronic African food crisis.

To address the soil fertility constraints, some global programmes have been initiated as soil health issues rose within the agendas of policymakers, development partners and donor agencies. The Soil Fertility Initiative was launched at the World Food Summit in 1996 and from 1998 to 2001, the Food and Agriculture Organization of the United Nations (FAO), International Centre for Research in Agroforestry (ICRAF), International Food Policy Research Institute (IFPRI), International Fertilizer Development Centre (IFDC), USAID, and the World Bank conducted consultations in 20 African countries and developed national action plans for soil fertility. The Abuja declaration, following from the African Fertilizer Summit of 2006 set the scene for major investments in boosting fertilizer supplies. CAADP – the Comprehensive African Agricultural Development Programme – has been active in supporting the follow-up to the summit, particularly through its work on improving markets and trade. AGRA – the Alliance for a Green Revolution in Africa – has launched a major new Soil Health Programme (SHP) aimed at 4.1 million farmers across Africa. Other initiatives abound – the Millennium Villages programme,

Sasakawa-Global 2000, the activities of the Association for Better Land Husbandry, among many others. All see soil fertility as central, although the suggested solutions and policy requirements are very different. Collective action is required by all actors along the production to marketing value chain to address the challenges of soil fertility in Africa.

Technologies Are Available

Several technological breakthroughs have emerged in Africa over the past decade that, once effectively disseminated, offer the means to reverse the land degradation situation. Long term experiments (LTE) have played a key role in understanding the changes in soil fertility resulting from changes in land management practices. Technologies that sustain increased crop yields, the result of researches over several seasons in different agro-climatic zones (AEZs), can be summarized into (1) inorganic–organic nutrient combinations and (2) nutrient and water combinations. These contribute to controlling nutrient mining and improve water and nutrient use efficiency, a key to higher agricultural productivity in the African continent. No doubt, fertilizers application is an important input in most cases, with P and N being the most limiting nutrients (Bationo et al. 1987).

Inorganic–Organic Nutrient Combinations

Research findings across diverse AEZs of sub-Saharan Africa show that the highest and most sustainable gains in crop productivity per unit nutrient are achieved when fertilizer and organic inputs are used in combination (Giller et al. 1998; Vanlauwe et al. 2001a). Combining the strategic application of chemical fertilizers and farmer-available organic resources increases nutrient use efficiency, makes fertilizer use more profitable and protects soil quality. Alone, application of mineral fertilizers on impoverished soils leads to positive crop yield responses but results from long-term experiments indicate that yields decline following continuous application of only mineral fertilizer. Such declines might result from (1) soil acidification by the fertilizers, (2) mining of nutrients as higher grain and straw yields remove more/ many nutrients than were added, (3) increased loss of nutrients through leaching as a result of the downward flux of nitrate when fertilizer N is added, and (4) decline of soil organic matter (SOM).

On the other hand, application of only organic inputs either as animal manure or plant residues decreases yields in many cases, and the application of organic materials is insufficient to meet the crop requirements for large scale food production. The combined use of mineral fertilizers and organic inputs, either as animal manures, compost, crop residue or agroforestry biomass, increases and maintains stable yields for extended periods, pointing to the need to integrate both the mineral and organic

Table 11.1 Effect of fertilizer and crop residue on pearl millet yield at Sadore, 2005 rainy season

Treatment	Grain yield (kg/ha)	TDM (kg/ha)
1 = Control	274	1,858
2 = Crop residue (CR)	374	2,619
3 = Fertilizer	903	4,013
4 = CR + F	1,309	5,580
SE	153	379
CV	43%	21%

fertilizers in crop production. For example, in an experiment established since 1986 in Niger, the traditional farmers' practices yielded only 25 kg/ha of pearl millet grain in 2005 whereas with application of 13 kg P/ha, 30 kg N/ha and crop residue in pearl millet following cowpea yielded 798 kg/ha of pearl millet grain (data not shown). In a trial established since 1982, sole application of CR increased the millet grain yield by 36% over a no-input control; this yield increased by a further 141% with fertilizer (F) application and by a further 45% when both CR and fertilizer were applied (CR + F; Table 11.1, see also Bationo et al. 2006).

In a field trial in southern Togo, Vanlauwe et al. (2001b) showed positive interactions between urea fertilizer and green manure application: the combined application of 45 kg urea-N ha^{-1} and 45 kg green manure-N ha^{-1} resulted in a yield benefit of 0.7 t grains ha^{-1} compared to the application of either source alone. Combination of manure and inorganic fertilizer increased crop productivity in West Africa by up to two times that of manure applied alone (Bationo et al. 2006) and several authors have shown that crop yields from the nutrient poor African soils can be substantially enhanced through use of manure (McIntire et al. 1992; Bationo and Buerkert 2001). These data indicate a high potential to increase crop yields in African farming systems. The improved agronomic performance when mineral and organic nutrient sources are combined arise because the minor nutrients essential for crop growth are lacking in the common mineral fertilizers but available in organic resources, the combination enables supply of all nutrients in suitable quantities and proportions and results in a general improvement in structure and soil fertility status, improved nutrient retention, turnover and availability (Nziguheba et al. 2000), and the organic amendments counteract soil acidity and Al toxicity (Pypers et al. 2005).

Long Term Experiments (LTE) that integrate legumes (grain and herbaceous) also contribute to improved soil fertility through biological nitrogen fixation and higher productivity. In an experiment established since 1993 at the research station of ICRISAT Sahelian Center at Sadore, Niger, rotation of cowpea and pearl millet increased pearl millet yield in 2005 by 20% over continuous pearl millet system for both with and without crop residue addition. In another study, Bationo and Vlek (1998) found an increase in N -use efficiency from 20% in the continuous cultivation of pearl millet (Pennisetum glaucum L.) to 28% when pearl millet was rotated with cowpea, a demonstration that N use efficiency can be increased through rotation of cereals with legumes. Positive effects of rotations are due to the improvement of soil

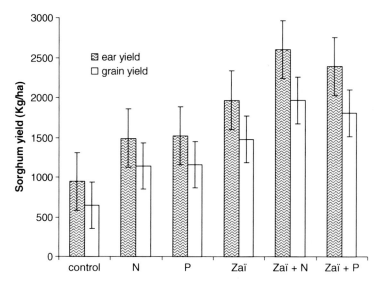

Fig. 11.1 Effect of water harvesting, N and P on sorghum yield in Tougouri, Burkina Faso, 2004–2005 season, error bars are LSD

biological and physical properties, the ability of some legumes to solubilize occluded P and highly insoluble calcium through legume root exudates (Arhara and Ohwaki 1989; Sahrawat et al. 2001), soil conservation, organic matter restoration and pest and disease control.

Water-Nutrient Combinations

Several technologies exist to improve water availability in drought-constrained areas, besides technologies improving the soil organic matter content. These technologies involve water harvesting through the use of "zai" pits, half moons, and stone bounds or tied ridging. Water harvesting strongly interacts with nutrient management. In Tougouri Burkina Faso, use of Zai, an insitu water harvesting technique, performed better than the use of either nitrogen or phosphorus fertilizer applied alone (Fig. 11.1); combination of zai with either N or P highly increased the yields indicating a better utilization of inorganic fertilizers with water harvesting. Other insitu water harvesting techniques such as half moon in West Africa and tied-ridges in eastern and southern Africa show that farmers can increase nutrient use efficiency, and both the yield of legume and cereal with water harvesting in the drier parts.

Despite the plausible technologies, yields in farmers field are several-fold lower than that in on-station experiments and in commercial farms. The lack of strategic planning and market development resources are identified as the factors impeding the widespread adoption of the proven land management practices and appropriate soil fertility products. It is believed that roughly half of the huge yield gap existing between SSA countries and the developed world will be closed through soil nutrients and improved agricultural practices; the other half through improved seed, but all this though strategic focus on the value addition and innovative financing.

Why Green Revolution (GR) Efforts Have Not Quite Succeeded

Good technologies and crop improvements by themselves are not enough to achieve the desired GR in Africa. Africa missed the green revolution of the 1960s and 1970s because, unlike Asia, (1) wheat and rice (the crops of the green revolution) are not the major food crops in Africa (where farming systems are dominated by root and tuber crops and sorghum and millet and other crops, including maize and pulses, crops that are more difficult to improve than wheat and rice), (2) Africa has more diverse agro-ecologies than Asia, so one-size-fits all technical change is impossible to achieve and (3) African agriculture is dominated by rain fed systems, with less than 5% of the cultivated land irrigated compared to over 45% of irrigated arable land in Asia. The effect of macroeconomic reforms such as removal of the government from the agricultural markets and the elimination of subsidies as part of the structural adjustment program initiated by the IMF and World bank in the 1980s decreased farmer access to credit at affordable rates ultimately leading to dramatic reduction in the adoption of modern crop varieties and fertilizers for a proportion of farmers. The removal of the government from agricultural markets has led to the entry of the private sector, but the markets for staple crops are still poorly organized, remain uncoordinated, have excessive transaction costs and risks, and are subject to price volatility which negatively affects net-buyers of food. To the contrary of high input prices, prices for agricultural produce have remained low.

Complementary investments in irrigation, roads, education, irrigation, fertilizers, energy and credit necessary for take-off of green revolution have been low and in some cases non-existent. Investment in rural roads, for example, leads to a marked reduction in marketing costs and margins and an increase in farm gate output prices but African governments are under-investing in such infrastructure. The road density in Africa has remained several magnitudes below what Asia had when it had the green revolution (Spencer 1994). Thus lack of infrastructure (roads, rails, ports, electricity and irrigation) continues to hamper the development of national agriculture, limits opportunities for intra-regional trade and lowers competitiveness in international markets.

The success with the earlier green revolutions in Zimbabwe (Rhodesia) and Kenya, although un-sustained, was linked to the existence of high caliber of scientists supported by their national governments' commitment to invest in long term research. Donor funding has fallen by at least 50% in the last 10 years, while national government expenditure for research declined from 0.8% of the agricultural GDP in 1980 to 0.3% in 1990s. The expenditure per scientist declined by 34% during 1961–1999; leading to the exodus of well trained staff from national programs (Haggblade 2005).

For a long time, the share of national budget to agriculture has been very low but this is changing with African governments' commitment to increase the allocation to at least 10% of their respective national budgets.

The use of agricultural inputs has remained low in Africa for decades generating a global irony: whereas the over-application of inorganic and organic fertilizers has led to environmental contamination in a number of areas in the developed world, insufficient application of nutrients and poor soil management, along with harsh climatic conditions and other factors, have contributed to the degradation of soils in sub-Saharan Africa (SSA). The average intensity of fertilizer use in SSA, excluding South Africa, is about 9 kg/ha. Intensity has generally been highest in Southern (16 kg/ha) and Eastern (8 kg/ha) Africa countries and lowest in the Sudano Sahel (4 kg/ha) and Central Africa (3 kg/ha). The diagnostic studies of fertilizer use in Africa have suggested that fertilizer use is low in Africa for four interrelated reasons:

• The low returns to fertilizer use due to agro-climatic conditions and current farming methods;
• The lack of information about fertilizer among retailers, farmers, and extension agents such as price information and best practices;
• The high costs of fertilizers due to increased agricultural commodity prices following a significant increase in demand arising from China, India and other emerging economies (FAO 2008), increase in energy prices for fertilizer production and shipping and inland transport, increase in mineral prices e.g., rock phosphate, and marginal expansion in fertilizer production,; and
• The inconsistent and adverse policy environment such as shifting government and donor subsidy policies that undermine private investment.

The recent increase in fertilizer prices raise concerns of its access and affordability to farmers. Since 2007 the nominal price of DAP increased from $200 to around $1,200 per metric ton – representing around a 500% increase in under 2 years; similar increases in the prices of both Urea and Rock phosphate have been observed. But while improving access to fertilizers is a necessary countermeasure, the low returns from unskilled use of these products present a major impediment to their adoption by most small-scale farmers. Training of agro-dealers and extension agents is thus important to ensure appropriate use of fertilizer.

Scaling up techniques have been flowed. Even though a wealth of knowledge and technology exists, most of these have not gone beyond the areas where they were developed and thus not readily available to small-scale farmers. Many farmers and farm households especially in the remote rural areas are not aware of available technologies and improved farm practices due to weakened public extension services. Other systems such as the farmer field schools mainly the work of NGOs have worked but they usually reach only a small fraction of farmers (mainly those in the limited project sites) and thus challenges still remain on how to scale-up such approaches to reach millions of farmers. Natural expansion of good practices has been limited by the lack of appropriate markets to catalyze it.

Try-it-yourself-to see-and-belief approaches have been applauded for the increased farmer adoption of ISFM technologies. The concept of "Farmers field School" (FFS) and demonstration trials for example ensures that knowledge and ideas are shared with the community through feedback mechanisms and are sustainable processes for farmers' capacity building but do not lead to sustained adoption of ISFM technologies. Although good results are achieved when farmers select, test and evaluate the performance of novel technologies at their farms, and study the how and why on a particular issue through programs such as the FFS, they need continued access to inputs and markets for their agricultural produce in order to sustain adoption of the appropriate technologies. Such input access and product markets are improved through appropriate and strategic financing and value addition at different stages of the input–output product chain.

New Challenges to the Achievement of a GR in Africa

Climate Change represents a formidable challenge that could exacerbate Africa's vulnerability to natural disasters and pose a huge challenge to increasing productivity. The projected impacts of climate change include greater frequency and severity of natural disasters in Africa. It is estimated that, without counter measures, rising temperatures resulting from climate change could lead to a 35% reduction in the productivity of African agriculture. Soil moisture stress, an overriding constraint to food production in much of Africa (only 14% of Africa is relatively free of moisture stress) is predicted to become more prevalent. Predictions of the International Panel on Climate Change (IPCC) suggest that Africa will experience increased drought in many areas in the near future. This is expected to increase the risk of investment in productivity enhancing technologies, unless appropriate water control measures are adopted. The increased risks further threaten an already weak agricultural sector credit availability, requiring mechanisms to guarantee credit providers.

The growing world fuel crisis is increasing the use of food grain in fuel production, with accelerating competition for agricultural land between biofuel and food crops. For example, a Norwegian firm has secured 38,000 ha of land for bio-fuel production in Ghana, and negotiations for land are on-going in several other African countries.

Paradigm Shifts in Soil Fertility Management

Over the years, the paradigms underlying soil fertility management research and development efforts have undergone substantial changes because of experiences gained with specific approaches and changes in the overall social, economic and political environment. During the 1960s and 1970s, an external input paradigm was driving the research and development agenda. The appropriate use of external inputs, be it fertilizers, lime, or irrigation water, was believed to be able to alleviate any constraint to crop production. Organic resources were considered less essential. Following this paradigm together with the use of improved cereal germplasm, the 'Green Revolution' boosted agricultural production in Asia and Latin America in ways not seen before. However, application of the 'Green Revolution' strategy in sub-Saharan Africa (SSA) resulted only in minor achievements because of a variety of reasons including the diversity of the agro-ecologies and cropping systems, variability in fertility, weak institutional arrangements, and lack of enabling policy.

In the early 1980s, the balance shifted from mineral inputs only, to low mineral input sustainable agriculture (LISA) where organic resources were believed to enable sustainable agricultural production. After a number of years of investment in research activities evaluating the potential of LISA technologies, several constraints were identified both at the technical (e.g., lack of sufficient organic resources) and the socio-economic level (e.g., labour intensive technologies). This led to the Second Paradigm for tropical soil fertility that recognized the need for both mineral and organic inputs to sustain crop production, and emphasized the need for all inputs to be used efficiently (Sanchez 1994). But although combining mineral and organic inputs resulted in greater benefits than either input alone, adoption of the 'Second Paradigm' by farmers was limited by the excessive requirement for land and labor to produce and process organic resources. The early 1980s and 1990s led to the emergence of the Integrated Natural Resource Management (INRM) research approach and ultimately the Integrated Soil Fertility Management (ISFM) paradigm. Although technically ISFM adopts the Second Paradigm, it recognizes the important role of social, cultural, and economic processes regulating soil fertility management strategies. Complete ISFM comprises the use of improved germplasm, fertilizer, appropriate organic resource management and adaptations to local conditions and seasonal events. Embedded within the ISFM paradigm is the Market-Led Integration Hypothesis which states that "improved profitability and access to market will motivate farmers to invest in new technology, particularly the integration of new varieties with improved soil management options". The hypothesis is based in part upon the disappointing past experiences of developing and promoting seemingly appropriate food production technologies, only to have them rejected by poor, risk-averse farmers unable or unwilling to invest in additional inputs. For example, in a study conducted in Vihiga and Kakamega Districts in western Kenya, 75% of farmers preferred 30 kg ha^{-1} P$_2$O$_5$ + 2.5 tha^{-1} FYM, despite not generating as high agronomic and economic returns as 60 P$_2$O$_5$ and 60 N kg ha^{-1}, but because they perceived that they could afford or access the requisite inputs. In Niger, hill placement of small quantities of P fertilizers was very attractive to the poor small-holder farmers. However, making the necessary investments in soil fertilization to derive benefits including adequate returns on investments depends on output markets and the market value of farm products and this varies across Africa, within regions and even within villages. The soil fertility restoration paradigms have failed to address constraints that are beyond farmer control. While the Market-led approach adopted in ISFM paradigm is plausible, parallel efforts to address the financing constraints that limit availability of agricultural inputs in rural agro-dealer shops have largely been lacking.

AGRA's Approach

AGRA aims 'to trigger (catalyze and bring about) an African led green revolution that will transform African agriculture into a highly productive, efficient, competitive and sustainable system that assures food security and lifts millions out of poverty'. AGRA uses fertilizer as the point of entry for its SHP and supports wholesale and retail fertilizer distribution channels at national and local levels, promotes fertilizer use and soil management at large scale for successful productivity and income growth in specific cropping systems, advocates and provides knowledge and technical support for policy changes that improve fertilizer procurement, support markets, and get knowledge to farmers; and improves technologies and data resources for soil health management and train and network the next generation of soil scientists. The program aims to (1) increase farmers' financial and physical access to the locally appropriate soil nutrients and fertilizers in an efficient, equitable and sustainable manner, (2) create access to locally appropriate ISFM knowledge, agronomic practice and technology packages and (3) create a national policy environment for investment in fertilizer and ISFM. For example to increase access and affordability of agro-inputs, AGRA is specifically supporting local blending and small packaging, supporting wholesale, retail and co-operative networks to increase input–output distribution, facilitating creation of a fertilizer market information and trading system, and improving African input procurement and production capacity. The result is increased fertilizer consumption, which has been shown to correlate positively with GDP.

The AGRA approach recognises the need to bridge the financial constraints and risks associated with farming in Africa as an important component to achieve sustained agricultural productivity and economic growth. An example is the provision of credit guarantees to banks to support input wholesalers and agro-dealers access to credit or working capital under improved terms as well as enabling farmer access to finance for the purchase of ISFM technology package inputs.

The New Paradigm

Quite often, farmers do not access the necessary services for increased and sustained agricultural productivity because of ill-equipped or non-existent dealers within the agricultural input and output value-chain continuum. For example, agro-dealers who stock essential farm supplies, such as seeds and fertilizers needed by farmers to increase yield and end poverty, do not have the necessary capital, and providers of such capital (mainly banks and microfinance institutions) fear the investment due to associated risks. Rather than the credit providers solely bearing the risks, mechanisms are needed for sharing that risk among several development partners, a move that obviously increases the investments in the risky agricultural sector. As discussed earlier, the new agricultural environment requires increased crop insurance and cushioning of investors in the agricultural value chain continuum especially in Africa. Innovative financing is the missing link/gap to support a revolution in small-holder agriculture. Innovative financing through microfinance institutions, banks and cooperatives could for instance provide opportunities for the agrodealers and enterprising farmers to access credit guarantees for increased investments. With such financing, an increasing proportion of poor African farmers can access and use adequate quantities of improved seeds and fertilizers at affordable prices. An example of such financing is AGRA's guarantee fund which provides the 'security' needed by the banks in order to give credit to partners in the agricultural sector who often have no collateral to pledge for repayment of loans. In its essence, innovative financing complements ISFM by addressing the missing financing gaps through strategic focus into the whole agricultural input–output spectrum. Thus, innovative financing broadens the existing ISFM paradigm into a new paradigm for improvement of soil fertility and crop productivity, and is termed the ISFM-IF paradigm.

The evidence of a green revolution through strategic (indirect) financing is the case of Malawi where smart subsidies have helped to increase farmers' access to inputs, moving the country from a net-food importer to exporter within few years. The Malawian government uses voucher-based smart subsidy scheme. Through the use of smart subsidies, the prices of fertilizers and other agricultural inputs are subsidized resulting to their increased affordability and use. The use of smart subsidies offers opportunity of improving farmer access to inputs by lowering the price or improving the availability of fertilizer to farmers in ways that encourage its efficient use and stimulate private input market development. Key characteristics of market-smart subsidies are that they do not distort the price of fertilizer relative to other inputs so that farmers use fertilizer in an economically efficient manner and, secondly, incentives for farmers and input suppliers are shifted in a way that contributes to a strengthening and deepening of the private input supply system in the long-term. Besides Malawi, there are indications of success following the use of such subsidies in Nigeria and several other countries in Africa (Kenya, Tanzania, Rwanda, Uganda, Ghana, and Mozambique) have started subsidy programs for their small-holder farmers. Optimal design of these programs is needed to ensure impact and reduce excessive cost and chance of corruption, and eventual end of subsidies are needed. Sustained demand for the generated agricultural produce is critical for continued adoption of the ISFM technology package, and this calls for investments in the output value addition and marketing.

A credit guarantee model, the Warrantage, operated at the village level in Niger, and later scaled up to Burkina Faso and Mali has demonstrated the potential to remove barriers to the adoption of soil fertility restoration and promote the economy of farming households (Tabo et al. 2007). Quite often, and in many parts of Africa, pressing demand for cash pushes poor farmers to sell off crop produce immediately after harvest when prices are very low. In the warrantage model, such farmers are provided with post-harvest credit on the basis of storage of grain as collateral ("warrantage"). The stored produce, characteristically grain, is sold later in the season at higher prices (up to 40% higher) and higher profits, thus increasing farmers' ability to buy external inputs (fertilizer and pesticides). In just 3 years for example, a total of about 5,000 "warrantage" farm households in 20 pilot sites in Niger started using better natural resource management technologies and are producing 50% more food. Also, in participating villages, farmers applying fertilizers (in microdoses (4 kg P/ha)), drastically increased, grain yields of millet and sorghum increased by between 43% and 120% and profits increased by 52–134% (Tabo et al. 2007). The "warrantage" participating farmers are organized into farmer-based enterprises or cooperative organizations that provide access to the micro-credit and inputs, as well as better access to output markets. The model adapts a 'FAO-led Projet Intrant' approach that promoted a system of credit adapted to the socio-economic conditions of the rural areas, which linked the requirement for guaranteed credit to the necessity of adding value to the agricultural products while organizing producers for the supply of inputs. This model aims at solving farmers' liquidity constraints through development of complementary institutional and market linkages and has enhanced the adoption of the fertilizer micro-dosing technology in West Africa. For its effective implementation, besides the sustainable farmer-based enterprises and cooperative organizations, storage facilities and inputs shops (boutique d' intrants) are built, and credit and savings schemes are also developed, all managed by members of these cooperatives. This practice, also known as the Inventory Credit System, has been practiced in Asia for many years but was recently introduced in West Africa where it has started already to make a difference in the livelihood of the rural communities. The warrantage models is a successful experience showing that supportive and complementary institutional innovation, market linkage and financing are required for the adoption of soil fertility improvement practices such as micro-dose technology.

Innovative financing in the form of loan guarantees provided to commercial banks allow target borrowers to access bank credit to enhance their operations within the agricultural input–output continuum. Such loan guarantees, established by AGRA and IFAD and other partners, are leveraging funds with banks such as Standard Charted Bank (Tanzania, Ghana, Mozambique, Uganda), Equity Bank-Kenya, National bank- Tanzania. This is targeted to remove the finance access problem that is a key constraint to private sector input distribution in many African countries. For example, a $5 m credit guarantee to Equity Bank in Kenya is increasing western Kenya wholesalers and importers access to credit to increase product supply, and allowing these borrowers time to sell the product in the market. Initiatives on agrodealer development are being implemented by organizations such as CNFA and IFDC through support from AGRA and other organizations. With business support to increase access to working capital, improve marketing of farm inputs and basic record keeping, agrodealers are becoming the private sector entities that are the smallholder's source for a range of inputs.

An African green revolution requires drastic change in policies affecting the agricultural sector and which influence ISFM adoption. Such policies (and institutional arrangements) are appropriate interventions that minimize distortions to farmers' and agrodealers' incentives and can thus support the development of input–output supply chains, ISFM knowledge and technology dissemination systems and improve soil water and nutrient conservation. It also requires development of complementary infrastructures such as roads that hasten input–output transportation. Thus nations that have or are investing in appropriate policies and institutional frameworks and complementary agricultural infrastructures are well placed to achieve agricultural green revolution than are those to the contrary. Investments towards a GR such as the AGRA program focus their investments in those countries where such requisite policies and institutional frameworks are available, or seem to be underway.

Conclusions and Way Forward

Investments to fill the existing financing gaps are needed to support sustained adoption and practice of sound land management practices by smallholder farmers. It is critically important to empower farmers in basic principles of crop nutrition and management since simple techniques such as correct planning distance and more precise fertilizer placement can double crop yields. Since Africa's staple crops are varied and are cultivated in highly diverse ecological zones, it is important that farmer education follows adaptation to local conditions and recommended practices are appropriate to farmers' choices in that locality. This calls for the existence of sound and functional technical support from the extension systems, both public and private.

High rates of fertilizer have long been recommended, but the systems to supply them are not in place. African markets are small and the prices charged by multinational fertilizer suppliers for the small lots imported into African countries are high. Domestic production is small. Transportation costs from ports to the interior are high and other marketing charges, including profits to wholesalers and retailers, are high. Farmers often end up paying double or triple the import price. There are relatively few dealers selling agricultural inputs to smallholders, they have limited capital, little ability to extend credit, and limited business acumen. Farm prices of fertilizer in Africa can be reduced by about 15% through country-specific strategies combining improved procurement, improved retail networks, reduced tariffs, disseminating competitive price information, local blending, and logistics coordination.

African countries could benefit from the more favourable economies of scale by establishing functional regional fertilizer procurement facilities. This will improve procurement systems (bulk importation) and transportation. Additional savings can be achieved by local bagging, local blending of imported materials, and eventually, local granulation. The African Development Bank (AfDB) is currently spearheading the establishment of an African Fertilizer Financing Mechanism (AFFM) to help countries access fertilizers at competitive prices.

Agrodealers are critical to farmers' access to affordable quantities of appropriate fertilizer in their local environments. With business support to increase access to working capital, improve marketing of farm inputs and basic record keeping, agrodealers are becoming the private sector entities that are the smallholder's source for a range of inputs. Innovative financing through microfinance institutions and banks could provide opportunities for the agrodealers and enterprising farmers to access credit guarantees for increased investments. Initiatives on agrodealer development are being implemented by organizations such as CNFA and IFDC through support from AGRA and other organizations. Components on innovative financing have also been established by AGRA and IFAD and other partners for leveraging funds with banks such as Standard Charted Bank (Tanzania, Ghana, Mozambique, Uganda), Equity Bank- Kenya, National bank-Tanzania.

Africa is endowed with numerous phosphate ore deposits, which are a potential source of phosphate fertilizers . However, few of these deposits have been developed, mainly due to their size and quality, limited domestic markets and depressed phosphate prices in the global fertilizer market, which do not justify investments and operating costs. Catalytic support is needed for private sector led nutrient/ supplement mining, fertilizer manufacturing and blending investments, through supporting development of potential economically viable opportunities to reduce cost of production of local nutrients and amendments and/or production of more locally appropriate fertilizer blends.

The use of smart subsidies offers opportunity of improving farmer access to inputs for increased agricultural production. Smart subsidies lower the price or improve the availability of fertilizer to farmers in ways that encourage its efficient use and stimulate private input market development. Key characteristics of market-smart subsidies are that they do not distort the price of fertilizer relative to other inputs so that farmers will use fertilizer in an economically efficient manner and, secondly, that incentives for farmers and input suppliers are shifted in a way that will contribute to a strengthening and deepening of the private input supply system in the long-term. There are indications of success from experience in Malawi, Nigeria, and several other countries in Africa (Kenya, Tanzania, Rwanda, Uganda, Ghana, and Mozambique) that have started subsidy programs for their smallholder farmers. Optimal design of these programs is needed to ensure impact and reduce excessive cost and chance of corruption, and eventual end of subsidies are needed.

References

Arhara J, OhwakiY (1989) Estimation of available phosphorus in vertisol and alfisol in view of root effects on rhizosphere soil. XI Colloquium, Wageningen, Holland

Bationo A, Buerkert A (2001) Soil organic carbon management for sustainable land use in Sudano-Sahelian West African. Nutr Cycl Agroecosyst 61:131–142

Bationo A, Vlek PLG (1998) The role of nitrogen fertilizers applied to food crops in the Sudano-Sahelian zone of West Africa. In: Renard G, Neef A, Becker K, von Oppen M (eds) Soil fertility management in West African land use systems. Margraf Verlag, Weikersheim, pp 41–51

Bationo A, Christianson CB, Mokwunye AU (1987) Soil fertility management of the millet-producing sandy soils of Sahelian West Africa: the Niger experience. Paper presented at the workshop on soil and crop management systems for rainfed agriculture in the Sudano-Sahelian zone, International Crops Research Institute for the Semi-Arid Tropics (ICRISAT), Niamey, Niger

Bationo A, Kihara J, Vanlauwe B, Kimetu J, Waswa B, Sahrawat KL (2006) Integrated nutrient management – Concepts and experience from sub-Saharan Africa. Hartworth Press, New York

FAO (2008) Current world fertilizer trends and outlook to 2011/12

FAO stat (2008) Food security statistics. Prevalence of undernourishment in total population. http://www.fao.org/economic/ess/food-security-statistics/en/. Accessed18 Aug 2009

Giller KE, Cadisch G, Mugwira LM (1998) Potential benefits from interactions between mineral and organic nutrient sources. In: Waddington SR, Murwira HK, Kumwenda JDT, Hikwa D, Tagwira F (eds) Soil fertility research for maize-based farming systems in Malawi and Zimbabwe. Soil Fertility Network and CIMMYT, Harare, pp 155–158

Haggblade S (2005) From roller coasters to rocket ships: the role of technology in African agricultural success. In: Djurfeldt G, Holmen H, Jirstrom M, Larsson R (eds) The African food crisis: lessons from the Asian green revolution. CABI, Wallingford/Cambridge

Maatman A, Wopereis MCS, Debrah KS, Groot JJR (2008) From thousands to millions: accelerating agricultural intensification and economic growth in sub-Saharan Africa. In: Bationo A, Waswa B, Kihara J, Kimetu J (eds) Advances in integrated soil fertility management in sub-Saharan Africa: challenges and opportunities. Springer, Dordrecht, p 1091

McIntire J, Bourzat D, Pingali P (1992) Crop-livestock interactions in Sub Saharan Africa. The World Bank, Washington, DC

Morris M, Kelly VA, Kopicki RJ, Byerlee D (2007) Fertilizer use in African agriculture. The World Bank, Washington, DC

Nziguheba G, Merckx R, Palm CA, Rao MR (2000) Organic residues affect phosphorus availability and maize yields in a nitisol of Western Kenya. Biol Fert Soils 32:328–339

Pypers P, Verstraete S, Thi CP, Merckx R (2005) Changes in mineral nitrogen, phosphorous availability and salt extractable aluminum following the application of green manure residues in two weathered soils of South Vietnam. Soil Biol Biochem 37:163–172

Sahrawat KL, Abekoe MK, Diatta S (2001) Application of inorganic phosphorus fertilizer. P. 225–246. In: Tian G, Ishida F, Keatinge D (eds) Sustaining soil fertility in West Africa. Soil Science Society of America special publication number 58. Soil Science Society of America and American Society of Agronomy, Madison

Sanchez PA (1994) Tropical soil fertility research toward the second paradigm. 15th World congress of soil science, Acapulco, Mexico, pp 65–88

Sanchez PA, Shepherd JD, Soule MJ, Place FM, Buresh RJ, Izac AMN, Mukwonye AU, Keswiga FR, Ndiritu CG, Woomer PL (1997) Soils fertility replenishment in Africa: an investment in natural resource capital. In: Buresh RJ, Sanchez PA, Calhoun F (eds) Replenishing soil fertility in Africa. Soil Science Society of America special publication number 51. Soil Science Society of America, Madison, pp 1–46

Smaling EMA, Stoorvogel JJ, Windmeijer PN (1993) Calculating soil nutrient balances in Africa at different scales II. District scale. Fert Res 35:237–350

Spencer DS (1994) Infrastructure and technology constraints to agricultural development in the humid and sub humid tropics of Africa. EPTD Discussion Paper No. 3, International Food Policy Research Institute, Washington, DC

Tabo R, Bationo A, Gerald B, Ndjeunga J, Marchal D, Amadou B, Annou MG, Sogodogo D, Taonda JBS, Hassane O, Diallo MK, Koala S (2007) Improving cereal productivity and farmers' income using a strategic application of fertilizers in West Africa. In: Bationo A, Waswa BS, Kihara J, Kimetu J (eds) Advances in integrated soil fertility management in Sub-Saharan Africa: challenges and opportunities. Springer, Dordrecht, pp 201–208

UN data (2006) World population prospects: the 2006 revision. http://data.un.org/Data.aspx?d=Po pDiv&f=variableID%3A47. Accessed 25 May 2009

Vanlauwe B, Aihou K, Aman S, Iwuafor ENO, Tossah BK, Diels J, Sanginga N, Merckx R, Deckers S (2001a) Maize yield as affected by organic inputs and Urea in the West-African moist Savanna. Agron J 93:1191–1199

Vanlauwe B, Wendt J, Diels J (2001a) Combined application of organic matter and fertilizer. In: Tian G, Ishida F, Keating JDH (eds) Sustaining soil fertility in West Africa. Soil Society of America special publication number 58. Soil Society of America, Madison, pp 247–280

World Bank (2009) "PovCalNet" database

Index

J. Kihara et al. (eds.), *Improving Soil Fertility Recommendations in Africa using
the Decision Support System for Agrotechnology Transfer (DSSAT)*,
DOI 10.1007/978-94-007-2960-5, © Springer Science+Business Media Dordrecht 2012

Printed by Publishers' Graphics LLC USA

2012